DOG TRAINING decoded

Simon Prins

First published in Germany in 2022 by Kynos Verlag Dr Dieter Fleig Gmbh

English language edition published in 2023 by First Stone Publishing,
an imprint of Westline Publishing Limited
The Old Hen House, St Martin's Farm, Zeals, BA12 6NZ, United Kingdom

ISBN 9781910488683

Printed through Printworks Global Ltd

1 2 3 4 5 6 7 8 9 0

ACKNOWLEDGEMENTS

If I had to list all the people who helped me along the way and got me to this point, I'll probably forget a few. There have been many great and generous minds to whom I owe much. Yes, I made big mistakes, spent countless hours training animals, read many books, articles and papers, missed a lot of sleep and had many interesting conversations with fellow trainers and scientists from around the world. A big thank you to all of you! And a big imaginary bone to all the dogs that gave me their trust and energy and helped me grow.

A special "thank you!" goes to Marian and Bob Bailey. They have been the most decisive door-openers for me. After our first meeting in the US we became close friends and worked closely together for more than 25 years. Before Marian passed, I was able to promise her that I would do everything in my power to spread her knowledge of the methods that had transformed animal training. In fact, I've had the privilege of training people all over the world and helping many talented dog trainers in police, customs, and special forces. I also got to know many highly motivated trainers from the civilian sports sector, got to work with zoo animal trainers and was fortunate to be able to work with very smart scientists on some very interesting topics. And I am thankful for all the dogs in my life, who have given me so much joy and confidence.

CONTENTS

FOREWORD

By Bob Bailey

It was 1997 when my home phone rang and the person on the line announced himself, saying: "Hi, I'm Simon Prins – I'm a police officer in Holland. I train police dogs, and I want to talk to you about what I do."

"Sorry, Simon," I replied quickly. "We don't work with policemen – thank you for calling." As far as I was concerned that was the end of the conversation, so I hung up.

Thirty years prior to this phone call, my wife, Marian, and our staff at Animal Behavior Enterprises (ABE), had trained many military and police dogs, and their handlers. While we had great success in training the dogs, the handlers were not receptive to our science (data) based approach to training. Later we discovered that many of the dogs we trained were given to inexperienced handlers and were used in ways of which we disapproved. We promised ourselves that we would never again train dogs for the police or the military, and we kept our word despite numerous requests from police units across the USA.

But Simon Prins proved to be unlike any of the policemen who had requested our help. First, he was persistent. He called several more times before we finally agreed to listen to what he had to say. He explained that he didn't like the way he was being told to train and handle dogs, and he was looking for a new direction. Following weeks of back-and-forth phone calls, Simon joined us, here in Hot Springs, with a handful of Dutch trainers from his police unit. The objective was to learn the principles of operant conditioning with chickens. I admit I was sceptical; I didn't think Simon would be able meet the demands I made on him and his colleagues. However, I was proved wrong. Simon changed his behaviour, quickly adapting to – and fulfilling – each of my 'demands'.

It is difficult to be objective when measuring the effectiveness of teaching an animal training programme. The human student trainer must learn to change many dimensions of their behaviour: what to reinforce, when to reinforce, when not to reinforce, when to move, when not to move, and much, much more. And all of these changes must take place immediately – in real time – not sometime later, after giving some thought to the matter. It is easier to measure the product of the student trainer's action – the animal's behaviour. The more 'wanted' behaviour we see, and the less 'unwanted' behaviour we see, shows us whether the animal's behaviour is 'better' or 'improved'. We measure the amount of wanted and unwanted behaviour by collecting and evaluating information, our observations, or 'data'.

Data was a new idea to most of the Dutch trainers. The exception was Simon. He had been collecting data for years, and he was already quite good at it, which was unusual for a trainer at that time. Training chickens to do simple behaviours is straightforward, but it is still not easy. The Dutch trainers were used to 'forceful' methods of changing behaviour. Chickens do not take kindly to this. After all, there is nothing keeping them on the table other than their desire to be there. It is, therefore, up to the trainer to give value to staying on the table and paying attention to the task at hand. This was a new idea to the student trainers. Some found it difficult to change their behaviour. Simon did not. The product of good training is a wanted change of the subject animal's behaviour. Simon's animals acquired the wanted behaviours. Not only did the chickens offer the wanted behaviours, over time they offered new behaviours faster and faster, the product of an improving trainer. I was impressed.

Impressive, too, was Simon's leadership, which was evident during training, and even while off-duty. The student-trainers were pleased to discover that Hot Springs had a fine German restaurant, the *Brau Haus*. Most evenings would find the fellows there, enjoying the food and the camaraderie. Even in this setting, Simon was a leader, and in a quiet way, his influence was obvious. Of course, influence can be of a coercive nature, i.e., being bossy. But, in my experience, the best teachers are less intrusive and have a more positive influence on others.

Over the next several years, even though on separate continents, Simon, Marian, and I worked together on developing training plans or protocols. We discussed the idea of behavioural systems, which included a trainer/handler, a dog, and often specialised devices for accomplishing tasks. Simon immediately saw the value of operational systems for police work, and by 2002 he was developing behavioural systems for his unit. This included the use of laser guidance and on-board cameras.

Over the last 25 years, Simon and I have worked together on numerous projects and, after the death of my wife Marian in 2001, I spent a lot of time in Europe, which distracted me from my loss. Simon and I spent many days together training, educating trainers, developing training plans and devising sophisticated training aids - and we had fun together. In 2019, I selected him as one of the two European representatives at my Hot Springs *Think, Plan, Do* *c*onference. I also consider Simon a good friend who shares with me joy and sorrow, success and failure. It's a privilege for me to be with him and his family - his wife, Jolanda, and his son, Rick.

Robert E. (Bob) Bailey, Sc.D.
Hot Springs, Arkansas.

PREFACE

Thank you for purchasing this book – you are now about to change the way you look at dog training. The only stipulation is that you need to be open-minded, curious and willing to change, because the magic will happen outside of your comfort zone!

I have been fortunate to travel extensively, train hundreds of different of dogs, and work with many incredible people. I have been continually challenged both by the dogs I have trained, and by attempting to fulfil the requirements of complex operational missions. It has

Simon Prins

Working with dogs and robots.

been a lifetime's experience in which success and failure often went hand in hand. Sometimes, like Don Quixote, I fought windmills – especially when I was desperate to replace punishment-based training with more positive methods. But once I had set out on my journey, there was no turning back.

Now I want to share my discoveries. There are a lot of books about dog training, but this one is unique! It's story is about bridging the gap between trainers, dog handlers and scientists. It's about real-life operations. It's about failure and success. It's about discovering the value of collecting and analysing data, and how to draw on scientific research. It is a manual on how to train dogs, but it will also give you new insights into training and behaviour and – perhaps our biggest challenge – it will help you to motivate trainers and handlers to work with positive reinforcement and to discard punishment as a means of changing behaviour.

Would you like to learn how to write your own training plans?

Would you like to understand how to collect and evaluate your training data?

Would you like to learn how to get other trainers to use more positive methods?

Do you want to understand the scientists?

Do you want the best for your dogs – and achieve optimum results?

Training is simple, but not easy. Once we stop judging ourselves, and judging others, we can start working together. We can learn faster and train our dogs more efficiently and effectively. This will lead to great training and, ultimately, to achieving better results. However, people are often averse to change and introducing the concept of operant conditioning, with the focus on positive reinforcement, into a world where punishment and coercion were the norm, was a huge challenge.

This was a time when police dog training was carried out secretly, behind closed doors. It was a random process with no prior knowledge of the outcome. Trainers just tried. Today knowledge is shared, and training has evolved into a structured process where we collect data, listen to scientists who educate handlers and trainers, and we take good care of our dogs. Why is this so important? Even in our technological age, with all the measuring devices, sensors and robots that we can put to use, dogs are still superior in so many ways.

Come with me on this journey of discovery – together we will find the answers to your questions and change your way of thinking. Throughout the process, our focus is to train in a more efficient and effective way, which is fun and enjoyable for the dogs – and also for the trainers!

Simon Prins
"Change something today and make tomorrow better…".

INTRODUCTION

From early childhood, I always pushed myself to the limit, both physically and mentally. My ambition was to be a professional diver, working on oil rigs, ships, bridges, and other large structures around the world. I had no thoughts of working with the police, let alone becoming an animal trainer. I signed up with and the Dutch Armed Forces and became the youngest student ever to graduate as a military diver. The training was tough and I learnt all about discipline, consequences, training programmes and training schedules.

Working as a young military NATO diver.

During my military service we did a lot of work for the Dutch National Police, and so I applied for a job with the police diving unit. I was invited to take the entrance exams and was judged suitable for the job. But then, to my surprise, I found out that the police did not have a full-time, dedicated diving unit. So I thanked them for the interest they had shown in me and prepared to leave. However, a member of the team persuaded me to stay, saying he was sure I would like working in the police force. As I had already quit my job in the military, I decided to give it a go – and I discovered I did really like it. The job was incredibly varied. After just a few months, I not only had a driver's licence for police vehicles and motorcycles, but also for high-speed police boats – and I was having a lot of fun.

It was during this time that I got involved in tracking down poachers and, as a result, I became interested in hunting dogs. I loved watching the hunters send their dogs to retrieve and was amazed at how easily they found both furred and feathered game, even in dense vegetation. I opted against hunting with a gun and became a gundog handler, working with Labrador Retrievers and Cocker Spaniels. I went out with the dogs almost every day and loved every minute of it. And when it got dark in the evening, I filled the hours discussing tactics with my hunting buddies, reading specialist books and watching videos.

The obvious next step was to become a police dog handler. With this in mind, I bought a Belgian Shepherd, a Malinois, and joined a local KPNV dog club. There are KNPV clubs all over the Netherlands where civilian handlers train their dogs in obedience, search work and protection, with the aim of preparing them for a role in the police. I started training my dog and soon discovered that use of punishment was widespread. Once I was accepted into the inner circle of the club, I saw even more examples of coercion. This was standard not only in my club, but in all KPNV clubs.

In the course of training, I was bitten several times by my own dog and, therefore, sought advice from the more experienced members of the club. They told me I had to *"break the dog's will"* and *"go alpha"*. I was young, I wanted to learn, and I imitated them. I started working with coercive methods and was successful, getting good results. However, it didn't feel right; I kept wondering if there might be a better way. Finally, I left the KNPV club because I could no longer tolerate their methods.

That same week I came across an advertisement asking for police dog handlers. I applied and was lucky; I was given the chance to join the dog unit. I was over the moon and told all my friends that I was now learning from the real pros. I will never forget the first day in my new job. No training programme, no books, no pdfs, no instructional videos. My instructor directed me to a squad car, and inside was a large, fierce-looking Malinois. He told me this was to be my police dog and I should get him out of the car. As I approached the vehicle, the dog raised his hackles and began to growl. He cocked his tail stiffly, his ears focused on me like radar screens, and his eyes flashed green. When I asked the instructor what I should do, he told me I had to quickly establish myself as an alpha and claim leadership as soon as I opened the crate. He instructed me to punish the dog and show him who was master – *"otherwise you'll end up in the hospital,"* he warned me. Wow! What an introduction to a world that I had revered for years. The so-called pros were no different from my KNPV clubmates, except that they wore uniforms. So, I opened the tailgate of the squad car, used all the coercion methods I had learnt, and started my career as a police dog handler in the worst possible way. Back then, the simple rule was:

If the dog does something you don't like, punish him.

If he does something you like, don't punish him.

And in that way, he learns what is required.

I was not allowed to use the food I had brought to the training session, play ball, or do anything else with the dog. My first 'training lessons' involved adherence to strict obedience, and punishment, and it was anything but fun – either for the dog or for me. But because of my lack of experience, I thought that, maybe, police dogs were different and I should, therefore, follow the 'official' training strategy. I had no idea how to achieve results without coercion, but I felt sure there had to be a better way than this rigorous, enforced training regime. I became determined to discover ways to train using more positive methods.

"First they ignore you. Then they laugh at you. Then they fight you. And then you win."

Mahatma Gandhi

In retrospect, the task I had set myself was anything but easy. Firstly, I had no knowledge of other training methods. Secondly, forced training produced results. Unwanted behaviour, such as getting bitten, disappeared and my dog showed the desired behaviour – *bite the suspect, not the handler* – that we needed in operational situations. Nevertheless, my gut feeling was that my dog and I could achieve much more if we understood each other better and could build trust in each other.

It may sound terrible, and I could be accused of being hypocritical, but I also learnt a lot from force-based training. It gets results and, in some situations, these are achieved more quickly compared with using positive reinforcement. If you have ever experienced being bitten by a police dog, I guarantee that you would use any form of training to stop it and prevent it from ever happening again. At least I can say that I understand the advantages and disadvantages of coercive training from my own, very concrete experience (including hospitalisation due to a serious bite injury). It made me appreciate how smart these animals are and, also, how much pressure they can take. I also came to understand why it is so difficult to talk to dog trainers about punishment and coercion, and why discussions become so heated. Shame and guilt – and how we deal with it – plays a big part. I have worked with compulsive coaches, positive coaches, and everyone in-between. I can speak to everyone without judgment; I understand their thinking. My mission is to open their eyes and help them discover new ways of thinking.

Many years have passed since I joined the police dog unit and, thankfully, there have been significant changes for the better. In the Netherlands, training methods have undergone a complete transformation. This is due, in part, to my work, but also to the many trainers and handlers in dog sports clubs, the police, military and customs dog squadrons who have contributed to changing the world of dog training. I was one of the coaches in this phase of change. I gathered knowledge, experience and trust, and we reached a point where the turnaround was irreversible. It didn't start on day one but, over time, I started to win the fight against the prevailing view that force-based training was the best way to train police dogs. People have called me a misfit, a rebel, a trickster, and a troublemaker. I take that as a positive because, at the end of the day, misfits are the people who are crazy enough to think they can change the world!

MY FIRST TEACHER

My first impetus for wanting change was Castor, my police dog. Yes, the one who bit me several times, the one who was considered difficult and dangerous, the one who nobody in the squadron wanted...

I go back to the time when I had been working with Castor for around six months. We were making some progress, but I was rebelling more and more against the force-based training that was imposed on us, so much so that I considered leaving the police dog unit. But what would become of Castor if I left? I decided I must learn more about animal training to see what might be possible. A landmark moment was reading *Don't Shoot the Dog!* by clicker training pioneer, Karen Pryor. I learnt about classical and operant conditioning steered by the work of Ivan Pavlov and B.F. Skinner and started experimenting with all kinds of reinforcements. The relationship with my dog became much closer. We understood each other better – and he stopped trying to bite me! I was allowed to take him home with me after work and he became part of my everyday life. We went on long hikes together and, yes, he saved me from several sticky situations while on duty. Our mutual respect grew. Then something happened on a police operation that changed everything.

It was on a Saturday evening. Our commander showed us a photo of a known criminal; he was wanted for rape and was on the run. The latest information was that he was still in town, and it was our job to find him and arrest him. A few hours later, while I was on patrol, I spotted him in the city centre. As an over-zealous young cop, I jumped out of the car, grabbed him, told him he was under arrest, and held him in a restrain. Not the best of plans, as it turned out. I was on my own, whereas he was accompanied by 10 buddies, who launched themselves into a full-scale attack on me. As I was being punched and kicked, I shouted for Castor, who was still in the car. We had practised this situation many times in training: the door of the crate is never locked so, when called, the dog jumps out of the car and bites the 'suspect' in the protective suit, who is pretending to attack the police officer. This was what I expected to happen. I called Castor and he responded instantly. He jumped out of the car and came within 5m of us - but he didn't bite! He growled and bared his teeth, causing my attackers to retreat. I continued to hold the wanted man while Castor pressed his back against my knee, making physical contact with me, and continuing to face the attackers. He was acting aggressively, but his behaviour was controlled. I dragged the suspect to the police car, handcuffed him, and got him into the car and, all the while, Castor maintained physical contact with me. Then he jumped back into his crate and we drove to the station to deposit the suspect in a cell.

What Castor had done for me was both unexpected and impressive. In the context of what was happening, his behaviour was faultless - but we had never trained this response. To this this day, I don't think I could set up a training plan to reproduce this behaviour. Despite all his previous training, which involved jumping out of the car and biting the attacker, Castor had chosen to behave differently - and he had selected the very best option for the situation. I no longer felt like the 'master' who had to use coercion to control a very aggressive police dog. We had become a team, built on friendship and trust.

That was the starting point for a change in my career as a police dog handler. Quite unexpectedly, I had been presented with proof that if the relationship between handler and dog is good, we should allow the dog to make decisions. Prior to this incident, my experience was that force-trained dogs functioned in a simple, analogous way. Decision-making was neither asked for, nor was it considered necessary. Basically, it was just a matter of *"attack!"* and *"let go!"*. But now I was developing a completely different relationship with my dog. It also seemed as though Castor was thinking more systematically, seeing the overall picture, and understanding that he could influence the situation.

As I write this, I am reminded of a statement made by Edward Thorndike, pioneer of operant conditioning:

"When an action leads to a reward, it is stamped into the mind of an animal."

Going back 30 years to the beginning of my career as a police dog handler, I was blind to the fact been that we must give animals a choice. Now I understand that they are not only committed physically, but also mentally. From that moment I decided to take a different path in training, moving away from compulsion and finding a new direction. The way forward was not always clear, and I experienced many ups and downs, but I never stopped learning.

Within a week of my decision to change course, the conflict became evident. In my small K9 unit, there were 10 dog handlers and one instructor. My instructor had a habit of correcting the dogs repeatedly, including my dog, Castor. Punishment was the only solution he knew, and he focused on wrong-doing, structuring training in such a way that the dogs were likely to make mistakes. Castor was a prime example. He always became very aggressive when he saw a protective suit or a bite sleeve and I struggled to prevent him from biting at will. The instructor kept on punishing him, but it made no difference. Now, with my change of heart, I was determined to put an end to this situation. This time when the instructor approached Castor, ready to punish him, I stopped him in his tracks. I told him that if he tried to hit my dog, I would let go of the leash, which meant result Castor would attack him. My colleagues froze; no one had ever dared to contradict the instructor. But I went a step further; I said that I would change the unwanted behaviour within two weeks without the use of punishment. He was furious! He yelled at me, saying that if there was one more incident of Castor biting without permission, I would be fired. Challenge accepted!

The training exercise we used to control bite behaviour was not complicated. Dog and handler were sent to search a building, room by room. The 'suspect', in a protective suit, choose a position which meant that the dog would encounter him while his handler was still outside the room. The challenge was for the dog to wait for the handler's command, rather than launching an attack. This was a real test as the dogs were trained to be fixated on the 'suspect', and the protective suit/bite sleeve was a real incentive. To prevent mistakes, we were taught to keep our dogs on a long leash; if we failed to stop an attack, the instructor would hit the dog and yell at us, which made these exercises truly nerve-wracking.

The dogs were so focused on biting the protective suit of the 'suspect', they often attacked without waiting for a cue from their handler.
Photo Adobe Stock.

I changed the rules. I recruited a friend to act as the 'suspect' and I practised the exercise several times a day in different environments. It was ok if Castor made mistakes – biting was no longer a penalty. Instead, the consequence of biting was that the 'suspect' (my friend) and the handler (me) became completely passive, and then started the exercise all over again. After about 10 restarts, Castor stopped attacking the 'suspect', sat down in front of him and barked – just as the textbook required! My response was to produce a ball on a rope and start a game. We had fun together! We kept working and, after a period of two weeks, the game had changed for Castor. He never again made the mistake of biting without a cue from me when faced with a 'suspect'. Instead, he sat down and barked. This was followed by one of four options:

- The 'suspect' rewarded Castor with a game of tug.
- I (the handler) rewarded him with a game of tug.
- I commanded Castor to come back to me, followed by a game of tug.
- The 'suspect' tried to run away, which was Castor's signal to attack – and then we brought him down together.

So, we had four options and a lot of opportunities for mistakes. But now mistakes no longer resulted in a penalty, but in a restart. We simply started the exercise all over again.

I will never forget the day when, two weeks later, my instructor asked me to give a demonstration of the exercise in front of my colleagues. Castor did a fantastic job! I explained my four options of reward, and everyone showed a real interest. But instead of the unit taking a different approach to training, the opposite happened. Although a few colleagues wanted to learn more about what I had been doing, most continued to work with coercion, as before. It was not until years later, when I became interested in human behaviour, that I realised they had simply remained within their comfort zone. But at the time, I could not understand why they were unwilling to shift their training to more positive, reward-based principles. People can change, but they don't want to be changed. Over the years I have learnt a lot about coaching people, too! Gradually, coercion was used less frequently in the unit, and a system of reward was increasingly preferred. However, the instructor was still against me and this did not create a good learning atmosphere. I was at the point of quitting when I got an unexpected phone call, asking me to join a newly formed police undercover unit.

Over the previous few months, a series of visitors had observed us at work. I had also had a couple of meetings with my commander, plus a number of seemingly random phone calls from people asking me questions about dog training. Then, I was invited to address a group of people who wanted to hear my opinion on the use of dogs in undercover operations. As a result, I was invited to work for the new unit. It would mean losing all the privileges I had acquired in my patrol unit, saying goodbye to my police dog, and leaving my colleagues without being able to tell them where I was going. But it was a tremendous challenge – and the team had done their homework – they knew I would not be able to resist their offer.

The year was 1996; it was the first day in my new job and I was sitting in an office, staring at a piece of paper, which listed a series of training objectives, such as: *Train a very quiet, passive drug detection dog that we can take with us on covert operations.*

At the time, drug detection dogs were known for their aggressive barking.

Train a dog that can follow a track in an urban environment.

Back then, tracking dogs only worked in rural areas.

The goal of the newly formed covert unit was to use dogs for tactical purposes. The management had brought in experts in audio, video, telemetry, sensors, and many other disciplines, all tasked with thinking outside the box and finding workable solutions. I had been brought in as the animal expert - at least that was what they thought they were getting. My brief was to introduce innovations in dog training - with no limits imposed. What a challenge!

In general, police dogs did not have a very good reputation in the force. They were deemed to be unreliable, smelly, noisy, limited in ability and the perpetrators of unwanted bite incidents. My job was to change all this and make dogs a useful and accepted tool in the toolbox of the new covert unit. The commander-in-chief was convinced that this could be achieved, but he knew we had to find a different way of working with dogs. He set me the task of analysing previous methodologies and changing them, where appropriate, so they were fit for purpose.

Among the many fascinating items on the list, there was one that really caught my attention:

Equip a dog with a camera system and other sensors and train him to go into an unknown, dangerous situation, out of sight of his handler.

This meant finding a way to covertly communicate with the dog over long distances, sending him into an operational area without prior training. I understood the essentials of disciplines such as searching for drugs, explosives, or people, but training a 'remote controlled' camera dog was something else. How do you learn about something that doesn't exist? A quote from actor/writer, Will Ferell, came to mind:

"The real way to learn anything is to get out there, try, and let your curiosity guide you."

THINKING AND PLANNING

The great thinkers of the past liked to drink absinthe, a spirit with a high alcohol content. It encouraged creativity and helped them find new ways of thinking. Great painters, such as Rembrandt, practised for hours on the smallest details to attain perfection. So, when faced with my training challenge, my plan was to think deeply, forget everything I had learnt up to that point, and decode animal training as if it was a computer programme.

I was accustomed to working through the steps of a protocol. As a diver in the military, this was the only way to survive in dangerous environments. Protocols are not set in stone; I see them as living documents that only get better with time and experience. We learn from our mistakes, at least when we have time for reflection. Unfortunately, too many people copy what others have done before. They switch off their own minds and follow a traditional method without trying to do better. I wanted to incorporate my military diving experience into my dog training plans. How could I do this? What would be different? What would be the same? I took out a large sheet of paper and started brainstorming by jotting down keywords in little clouds. No idea was too far-fetched and, after several hours and with help of a little whisky, I was done. I knew what I had to do and where to start.

The next morning, I asked to speak to my commander. I had a plan; it was not fully baked, but at least I had a framework. This was the beginning of my *think, plan, do* system of training. I explained to my commander that many of the items on the list were not hard to achieve. Of course, I could train a passive drug-sniffing dog, there would be no problem in training tracker dogs to work in different environments, nor would it be an issue to train a dog to respond to 'different' commands when involved in undercover operations. For example, I could train a dog that would run when I cued *"sit"* and would lie down when I cued *"here"* – a concept that proved highly successful in the months to come. However, I needed time, and a budget.

I needed money to buy dogs and, equally important, I needed the finances to fly around the world, interviewing leading trainers about their experiences, analysing their training plans, and training with them. I suggested a three-year timescale: the first year for research and building a directional training programme from scratch, the second year for training a variety of different dogs, and the third year for taking these dogs into operational situations to see how they would do. The commander stared at me when I asked for three years. He told me that money was not a problem, but three years was a very long time. I explained that my plan involved changing many traditional training methods, and we needed to give the project a fair chance. Following another eye-to-eye duel, he made me an offer:

"I'll give you your budget, and I'll give you three years to prove your work in the operational field. If I like it and you achieve your goals, you can stay here in the unit. If not, you go back to the general uniform patrol where you came from."

We shook hands. The date was April 1st, 1996 – and no, it was not a joke! The clock was ticking, and the project was underway.

THE PAVLOV PROJECT BEGINS

My starting point was training dogs in quiet, passive scent detection. I didn't think it would pose a major problem, and I was pretty sure I could learn the essentials at our national police dog school. I figured that training a dog, equipped with a camera and sensors, to work at a distance would probably be the most challenging. But, as it turned out, nothing proved to be easy.

Looking around, I saw a lot of very active scent detection dogs: barking, scratching, biting, or showing other forms of highly physical signalling. Their basic obedience was limited, and I felt they lacked in overall quality. I needed silent and passive scent detection dogs for our undercover investigations. They had to be reliable, obedient and sociable; they also needed to be mentally strong under pressure and remain calm for lengthy periods. When I started work at the police dog training school, I was surprised to discover that there were no training plans or handouts. Everything relied on oral tradition and craftsmanship, just as I had experienced when working patrol dogs. The instructors all had their own ideas, and no one would commit to a timescale for training a scent detection dog. I was told it was somewhere between three and nine months, without further explanation, Problems also arose with the 'man trailers', the single purpose tracking dogs. These dogs did a reasonable job in rural areas if no one else had walked over the tracks in the last 24 hours. However, once the ground became contaminated, problems arose, and following tracks on hard surfaces, such as tarmac, was deemed impossible.

My first camara dog, Andor, pictured in 1996.

I realised that three years would not be long enough. But there it was, the deadline had been set, and I was determined to succeed. I started by accompanying scent detection dogs and tracking dogs every day for the next six months. They trained mostly in the morning, so I would then spend the afternoons testing my ideas for training 'camera dogs', as I called them. I used my experience of training hunting dogs and combined it with what I had learnt in the police dog unit. I discovered that these three areas of training – scent detection, tracking and camera dogs – had one thing in common. Trust! Instructors who became interested in the camera dogs tried to help me, claiming that I had to use coercion. They believed that only a dog who had received this form of training would be 100 per cent reliable and would keep moving after receiving a command. Surprised, I asked what form the training should take. Whereupon the chief instructor outlined the following exercise:

"Take a long rope, about a 100m in length. Pass it over a pulley, mounted in a tree at a distance of some 50m, and then back on to the dog's collar. Then give the dog the command 'Forward!' and pull the rope. The dog can't help but give in to the pull and move towards the tree. This conditions him to move forward at the command, 'Forward!'.

As he walked away, he added:

"And use a prong collar!"

The flaws in his plan – quite apart from the use of coercion – made it a complete non-starter in a real-life situation. How would we cope in the middle of the night when it was dark, without a tree, a pulley or a 100m rope? I thanked him for his advice and duly ignored it. Instead, I continued to draw on my experiences from the world of hunting, but that wasn't enough either. In an operational situation, I would not be able to throw out dummies to motivate the dog to work. I was stuck! I needed a reliable training method, but despite searching worldwide, no one seemed to be involved in in this kind of work.

It was a Wednesday, and the scent detection dog handlers had an administrative meeting. I had decided to take the afternoon off and visit a nearby dolphinarium. When I was a child, I used to go there with my parents and had fond memories of the dolphin shows. Sitting in the audience, I was enjoying the show when, suddenly, I felt as though I had been struck by lightning. I was watching a young dolphin trainer standing on the podium, working with five dolphins at the same time. She was giving them directional cues and, all the while, she was smiling as she entertained the audience, which composed of some 500 people. She could do something that I had failed to do! I had problems with just one dog, on a long leash, when I tried to send him in various directions. The dolphin trainer was giving directional commands to five animals without using a training aid, or any form of compulsion. The dolphins' co-operation stemmed from trust in their trainer, and a sense of fun and enjoyment. I really needed to find out how these dolphins were trained!

The next morning at 10 am, the typical time for a coffee break in my country, I stood at the entrance to the dolphinarium. This time I wore my police uniform, brought a large cake and told the receptionist that I had an appointment with the head trainer. She made a phone call and 10 minutes later the training manager arrived. She introduced herself as Toinny and apologised, saying she must have forgotten the 'appointment'. We sat down, drank coffee, ate cake and ended up talking all day long. She showed me around, explained how the shows were set up and how the marine mammals (there were others besides dolphins) were trained.

The next weekend I paid a visit to my younger brother who had just entered the new world of the internet. Yes, younger people may not be able to imagine this, but in 1996 the internet was available for the first time! My brother had one of the earliest connections, and we spent the weekend finding out more about dolphin training. Eventually, we stumbled upon Keller and Marian Breland in the US who had been doing work with dolphins in the 1950s. We discovered that Keller was dead, but Marian was alive and married to a certain Bob Bailey. Both were very active in animal training based on operant conditioning. After a few days, we finally managed to find out their phone number in the US.

I was pretty nervous when I finally called the number and I was surprised when, after a few seconds, Bob Bailey answered in person. I introduced myself as a Dutch police officer tasked with a special project in dog training. I said I wanted to find out more about their work. Bob specifically asked if I was a police dog trainer, and when I said yes, he told me that, years ago, he had decided never to work with police dog trainers because of their coercive methods. He said goodbye, and just hung up. I was undaunted. I called again the next day - and at least 20 more times - to persuade him I was not a conventional police dog trainer. Finally, we got involved in a longer conversation and after consultation with his wife, Marian, they decided to give me a chance. If I came to the US, we could have dinner at a restaurant in Hot Springs, Arkansas, and the next day I would take a test. Hurray - I had an appointment!

A few weeks later I flew to the USA and met Marian and Bob in the evening at *Brau Haus*, a restaurant specialising in German cuisine. We had a long chat and huge schnitzels. At the end of the evening, they consulted briefly and then said they were ready to take the next step with me - the test - which was scheduled for the following morning. They gave me their address and told me to be there at 9 am sharp. When I entered the house, they led me to a long table in the living room, where sheets of paper were laid out. This was to be my training area. Bob then asked me to follow him into the garden and opened a small chicken coop. I had to take one of the chickens and bring it to the living room table. Marian gave me a clicker, a cup of chicken feed and a paper target. My test had begun. The objective was to demonstrate that I could train an animal – in this case a chicken – by either offering or withholding food, using the clicker as a bridge signal. They observed my progress, assessing my practical skills, my timing and my decision-making. I was way out of my comfort zone!

The training was divided into 'sessions' and 'trials'. After training different chickens for 45 minutes, Marian gave me a 45-minute theory lecture. Then, for the next 45 minutes, Bob showed me how I could improve my practical skills with the chickens, followed by another theory lesson with Marian. This was the routine I had to follow: training, theory, training, theory. It was hard work and when I drove back to my hotel that evening, I was very tired. My notebook was full of notes and jottings, and I still had homework to do – writing a training plan for the next day. Bob had already told me the objective – but not how to do it – that was my job to outline in a training plan. When I flew home two weeks later, I was exhausted but also very happy. I had learnt so much. With the help of Bob and Marian, I had trained chickens to peck at targets, identify colours, complete an obstacle course, pick up and pull things, and more. It was absolutely amazing! On the return flight to Amsterdam, the person seated next to me asked me whether I had been to the USA for vacation or work. I explained that I was a police officer and that it was a business trip. He enquired further, but I smiled to myself, thinking there was no way I could explain to him that I had been training chickens for the past two weeks. He would think I was crazy! I mumbled something about an exchange programme and then fell into a deep sleep.

Meeting Bob and Marian changed my life. I owe them so much. They opened my eyes and, without exaggeration, showed me a way out of the dark and into the light of animal training. They taught me how write and use training plans, collect and evaluate data, observe, segment and generalise behaviour. They gave advice on selecting and coaching trainers, and so much more. Marian died on September 25, 2001. Shortly before her death, I promised her that I would do everything in my power to bring the principles of operant conditioning to the world of police dogs. Bob and I have become good friends and, over the years, and we have worked on a wide range of training programmes. Together we have trained thousands of animal trainers around the world in workshops and seminars and put all our energy into achieving our goal – changing the world of professional dog training.

This book does not provide a *Simon Prins* training method. Equally, there is no such thing as a *Bob Bailey* training method. It's about classical and operant conditioning, and the great pioneers who developed this methodology. Yes, I created the training programmes for laser and radio-controlled camera police dogs, but I didn't put my name to them because it's just operant conditioning. Bob and Marian also sparked my interest in science, and, thanks to them, I have developed my knowledge by reading studies on animal behaviour and human psychology. I am also grateful to Bob for teaching me that time is precious – something I didn't really appreciate around the year 2000, when we were working so closely together. Now, more than 20 years later, I understand what he meant. The older we get, the faster time flies. Before we have even looked around, it will be Christmas and another year.

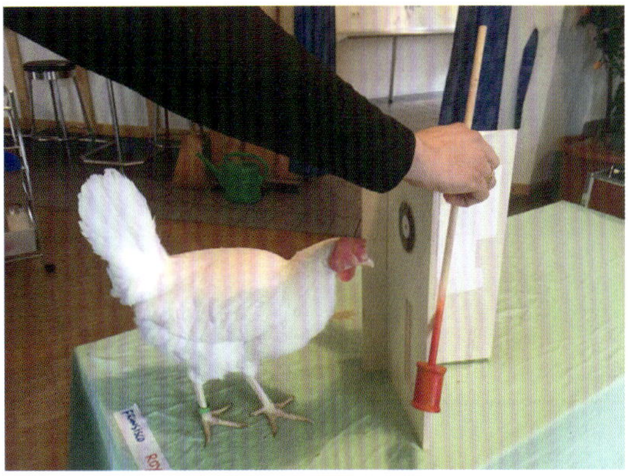

I learnt the principles of operant conditioning from Bob and Marian Bailey using chickens as my trainees.

For me, a particularly good example of the value of time is remembering our first steps in the remote camera dog training programme. As already outlined, the task was to train a dog, equipped with a camera, who could be controlled by his handler with directional cues at a distance of 200-300m, and this could be in the dark. The dog had to work quietly and in a covert way.

In the initial stages of training, I used a torch to illuminate the targets in the dark. Then, all of a sudden, it dawned on me: why not use a laser pointer? If I could figure out how to train the dog to approach a laser dot as a target, I could use this to give him directional cues. The drawback? At that time, a laser device cost £20,000! Now you can buy a laser pointer at any hardware store for £2. It took me nine months to condition a dog to approach a laser point from a distance of 150m, pick up an item, or look at a specific object so that the camera, fitted to his head, could film it. Now I can train this, and even more complex behaviours in less than four weeks. This is because I gathered so much data while I was training, and always worked with a plan. This was entirely thanks to Bob and Marian who taught me this, and so many other things, Most importantly:

- **Keep improving your training.**
- **Work towards becoming more effective and efficient.**
- **Change behaviour faster.**
- **Never stop learning!**

Of course, there were moments when I questioned the power of operant conditioning. I remember one of those incidents vividly. It was the first time I sent a camera dog, who was fitted with very expensive camera equipment, on a real-life mission. It was the middle of the night and I was crouched down, camouflaged in the bushes, waiting for my commander's signal to send the dog towards the target area, known as the 'hotspot'. A number of colleagues from our unit were present, watching this first operation with excitement. As soon as I received the signal to go, I sent my dog to the hotspot – in this case a building at a distance of some 200m – and tracked his movements with night-vision goggles. After about 50m, the dog suddenly turned left, away from the target, and accelerated. I was shocked! And then I also saw the reason for his change in behaviour: he was chasing a rabbit! We had to cancel the mission...

I was mortified, and I couldn't sleep that night. The next day I called Bob Bailey and told him that operant conditioning and positive reinforcement didn't work when it mattered. His laconic reply was: *"Well, Simon, try taking a big stick and beat the dog"*. With that he ended the conversation and left me hanging in mid-air. He didn't want me to slide back into using punishment. He wanted me to open my mind, to reflect on what had happened, to find a solution, and to change my training protocol. He didn't give me the solution; I had to find it myself which, from the point of view of a trainer, is much more sustainable. And yes, I found the solution – simply incorporating 'rabbits as a distraction' into my training plan.

By the time the three years of our project were up, we had all learnt a lot. The results exceeded our expectations; the dogs we had trained were being used in real-life missions with considerable success. Unnoticed by third parties, the camera dogs gave us insights into the terrain and buildings that were previously unavailable and, thus, provided us with information relevant to the operation. The scent detection dogs found things in places we couldn't even have imagined before the project started. The radio-controlled dogs placed information-gathering equipment in strategic places, and much more. We were also able to share our knowledge with other special police and military dog units around the world and this opened many doors for other colleagues, too.

Looking back today, I am very proud and grateful that I have been able to work with dogs in covert operations, in all kinds of disciplines. I know what it's like out there, working with a dog in dangerous situations. I have witnessed dogs saving the lives of colleagues and, on many occasions, I have seen our scent detection dogs finding evidence that human trackers had missed. I have been fortunate to work with many different dogs and people worldwide, and I have learnt from them all. However, the most important lesson I have learnt is that we can only make a difference if we stay open-minded, curious and humble; we must be willing to learn from each other and work together.

The improvised, heavy-duty radio and camera equipment used in the early days has now become a sophisticated high-tech system, which is used to communicate with dogs from a distance, and to see in real time what the dog sees.

Chapter 1:
MOVERS AND SHAKERS

Philosophers, psychologists, biologists, researchers, and other scientists have given us insights into how animals learn, how they use their senses, what motivates them, what scares them, how their memory works - and how we can train them better. The wonderful thing about our modern times is that we can take this learning further by using high-tech devices to measure brain activity in real time, which has given us some exciting information. However, we should not forget how much information is already out there. We need to take a look at what has already been discovered and, as an added bonus, to use modern technology to clarify earlier findings. We can learn so much from the pioneers who have conducted experiments, collected data, and developed training programmes - the results speak for themselves. Fortunately, aspects of this training have been documented, and data has been recorded.

In fact, our experience of training animals goes much further back - we have used animals for hunting, herding, guarding, and on the field of battle for thousands of years. As well as dogs performing war service, pigeons, marine mammals, rats, monkeys and bats, and even elephants - going back to the times of Hannibal - have been employed. Today dogs are trained in medical detection, for wildlife protection, to help people who are deaf or blind, and also those with physical disabilities. Throughout history there are stories of animals saving human lives, as well as using their skills to solve problems and to assist their human owners.

However, it is sad to note that, even after thousands of years' experience, we are still struggling to find the most effective and efficient way to train animals and educate animal trainers. My mission is to right this wrong, which I will outline in this book. But before we delve deeper into the world of dog training, let's take a quick look at the history of animal behaviour, as I believe we need to understand earlier concepts and ideas if we are to find a way forward.

In the early 19th century, psychologists were trying to find out more about human behaviour, but, as experiments in this field were not allowed, they turned their attention to animal behaviour, hoping their findings would shed light on the human condition. Thus, brilliant minds began to pave the way for modern psychology, and while their insights, theories, and principles have evolved, much of their original thinking is still relevant today.

EDWARD LEE THORNDIKE

(1874-1949), USA

Edward Lee Thorndike was a pioneer in the field of operant conditioning. He worked with 'puzzle boxes', the door of which could be opened from the inside by operating a simple mechanism. He locked a cat in the box, placed some tempting food in front of it, and noted the number of attempts, and the time it took for the cat to open the box and get the food. He concluded: *If an action leads to success, it is imprinted in the animal's memory*. He called this the *Law of Effect* or *Law of Effects*. So, it was Thorndike who showed us that an individual will show us more of a behaviour when it is followed by a positive consequence, and less when it is followed by a negative consequence. Thorndike also believed that intelligence should be viewed as a collection of different learnt skills rather than as an individual trait.

The law of effect
An individual will exhibit more of a behaviour when it is followed by a positive consequence and less when it is followed by a negative consequence.
Edward Lee Thorndike

JOHN BROADUS WATSON

(1878-1969), USA
Known as the father of behaviorism, Watson argued that it should be possible to measure behaviour objectively. He focused on observed behaviour and only used findings that were verifiable.

Watson conducted the infamous *Little Albert* experiment, named after a two-year-old boy who regularly played with his pet rat. One day, just as Albert was about to pet the rat, Watson struck a gong making an ear-splitting sound. Albert was startled and turned away from the rat. After repeating this process a few times, Albert was afraid of even looking at the rat, even though he no longer heard the sound of the gong. His fear of rats was conditioned; in addition, the sight of a white rabbit triggered his fear reaction. He later generalised his fear to dogs and then to all furry animals. Of course, this experiment provoked a discussion about what is deemed to be ethically permissible in human conditioning research.

Interestingly, Watson made a fortune from advertising. His psychological research improved the marketing success of many large companies, and we are still influenced by the results of his work.

Watson's finding
Conditioned fear responses can be generalised to occur in the presence other triggers.
John Broadus Watson

IVAN PETROVICH PAVLOV

(1849-1936), RUSSIA
At about the same time as Watson was conducting his research, a Russian psychologist was studying animal behaviour and, quite by accident, he stumbled upon what we now know as *classical conditioning*. Pavlov wanted to examine saliva so he gave meat powder to a number of test dogs and measured the amount of saliva that was produced. During the experiment, Pavlov noticed that the dogs began salivating even before the meat powder was given to them. Just looking at the scientists, in their white lab coats, was enough to start them salivating. This prompted him to run the famous experiment in which dogs were conditioned to salivate at the sound of the bell.

A neutral stimulus, such as a white coat or the sound of a bell, becomes a signal when followed by a primary reinforcer such as food. The result of this research has become known as the classical or Pavlovian conditioning, or respondent model (see *page 53*). This discovery has made Pavlov a key figure in modern, scientifically based, animal training. His influence is evident in all behavioural studies - a frequently heard quote among trainers is: "*Pavlov is always sitting on your shoulder*!"

Pavlov's discovery of classical conditioning

Dog hears bell and then gets food and hears bell - he salivates. Repeat often. Dog hears bell but gets no food - he salivates anyway.

Ivan Petrovich Pavlov

BURRHUS FREDERIC SKINNER

(1904-1990), USA

It was B.F. Skinner who brought the work of Thorndike, Watson and Pavlov together, outside the laboratory, and introduced it to a wider audience, earning the title, the father of operant conditioning. Skinner explained how we can influence and manipulate behavior through the use of reinforcement and punishment. Many modern trainers present Skinner's work in terms of the famous 'four quadrants', using the terms positive and negative reinforcement, and positive and negative punishment.

His famous book *The Behavior of Organisms* is a collection of documents on numerous experiments he carried out on animals. In his view, there are three responses that can follow a behaviour:

Neutral - the likelihood of the behaviour is neither increased nor decreased.

Reinforcer - the behaviour intensifies.

Punitive - the behaviour weakens.

Both reinforcers and punishers can be positive (added) or negative (taken away). Skinner was also a 'behaviourist' but, unlike Watson, he stressed that internal psychological triggers can also influence behaviour - although these can never be the ultimate cause of internal or external behaviour. Behaviour can never be the reason for behaviour. Behaviour is always the product of the interaction of an organism's history (genes or environment) and the circumstances or context in which it currently finds itself.

Skinner's three responses to behaviour

Neutral - The probability of the previous behaviour is neither increased nor decreased.

Reinforcer -The probability of the previous behaviour is increased.

Punisher - The probability of the previous behaviour is reduced.

Burrhus Frederic Skinner

In *The Behavior of Organisms*, Skinner describes his seminal work from 1938 in great detail. This is not easy reading so if you want to learn more about B.F. Skinner as a person and his life's work, I recommend Daniel W. Bjork's lively biography, *B.F. Skinner - A Life*, which provides fascinating insight into this remarkable personality.

KELLER BRELAND

(1915-1965), USA

MARIAN RUTH KRUSE

(1920-2001), USA

Keller Breland and Marian Ruth Kruse were two of Skinner's top students. After developing an understanding of the mysteries of animal behaviour, they unlocked its potential for animal training and discovered how powerful operant conditioning can be - not just in the lab, but in everyday life. During World War Two, they were the driving force behind the groundbreaking Pelican project which involved training pigeons to guide missiles. Although this experiment was never put into practice, they recognised the amazing power of operant conditioning.

Keller and Marian married on August 1, 1941, and as the war drew to a close, they bought a house, with some land, in Minnesota and began training animals. They founded a company called *Animal Behavior Enterprise (ABE)* and trained over 10,000 animals from over 140 different species, proving the importance and versatility of operant conditioning. They created the first dolphin shows, appeared in commercials featuring animals, assisted the US government in establishing military and security animal projects, developed training tools, and wrote numerous manuals and articles on animal training.

In 1950 *Animal Behavior Enterprise* moved to Hot Springs, Arkansas, and soon gained a worldwide reputation. In 1955 the Brelands opened their *IQ Zoo* where animals performed a variety of acts including dancing chickens, piano-playing ducks, air shows with free-flying birds, basketball-playing raccoons, and many other acts to entertain the audience. The *IQ Zoo* showcased the power of operant conditioning; the animals' learnt behaviour was strong and displayed fluently. No punishment was used – positive reinforcement was key. Not only did the Brelands generate income from the venture, but they were also able to fulfil their mission of bringing operant conditioning to a wider audience. They also took on the role of teaching trainers and gained an unparalleled reputation; an animal trainer who came from *ABE* was recognised as a top-class professional.

Behind the scenes, *ABE* has also been the base for government-funded undercover projects involving animals. This brought Bob Bailey to *ABE*. He was director of the US Navy marine mammal program, and he enlisted the Brelands to work with him on the project. The Brelands had also been working on a book, *Animal Behavior*, and when Keller Breland died on June 16, 1965, work was almost complete. Marian Breland published it in 1966 and, to this day, it remains a classic among animal trainers. Keller's death was an immense loss, but Marian Breland, Bob Bailey and the *ABE* team, continued to grow the company. Their work included top-secret research conducted on behalf of the military, the Secret Service, the CIA, and other agencies. Today many animal trainers, worldwide, are heavily influenced by what the Brelands achieved in the first half of the 20th century.

The Brelands' contribution:
The first trainers to put the principles of operant conditioning into practice and popularise the technique for a wider audience.

Keller Breland and Marian Ruth Kruse

BOB BAILEY
(b. 1936), USA
&
MARIAN BRELAND BAILEY
(1920-2001), USA

Even as a child, Bob was fascinated by animal behaviour and, as a teenager, he spent weeks in the desert near his home to see if he could influence coyote behaviour

Pictured left to right: Professor Skinner, Marian and Bob Bailey.

by feeding them in the same places, over and over again. He later went on to study astronomy, but then his earlier fascination with animal behaviour took over and he became a biologist. It was a landmark moment in his career when he was appointed executive director of the US Navy's dolphin programme. He had never trained a dolphin before, but he understood the importance of this government-run project, which eventually brought him into contact with the Brelands in 1955.

At that time there was only one commercially trained dolphin in the US, based at *Marine Studios* in Florida. But then disaster struck; the trainer quit his job, and the dolphin died. Keller and Marian Breland, whose work at the *IQ Zoo* has been featured in *Time Magazine*, were tasked with putting on a new show. Keller had never seen a dolphin in his life, but within six weeks the pair managed to stage a new dolphin show, featuring two dolphins and their trained helpers. It was easy for a biologist like Bob Bailey to set up an animal behaviour research programme, but training dolphins was something else entirely. When he heard about the Brelands' work at *Marine Studios*, he asked them to help with the Navy's dolphin project.

When the three got together, Bob saw with his own eyes what the Brelands had achieved at the *IQ Zoo;* he came into contact with the *ABE* trainers and became increasingly fascinated with animal training techniques. Long story short, with the Brelands' help, the Navy's dolphin programme was a huge success, and it marked the start of a close working relationship. The dolphin programme was the first of many government-run projects, with other animal species, that were conducted, either openly or secretly, at the *IQ Zoo. ABE* trainers were also hired to train government unit animal trainers in 1997.

When Keller Breland died, Bob and Marian fell in love, married in 1976, and had twins three times in a row! With their large family they lived and worked on the premises of the *IQ Zoo*, training animals and running courses for trainers. Government projects were expanding and my own research, which involved covert animal training programmes, was the starting point of my close working relationship and friendship with Bob and Marian, whom I first met in 1997.

Bob and Marian Bailey are known worldwide as the founders of the famous chicken camps, where students learnt about, understood, and applied operant conditioning. They discovered that chickens were best suited as learning models for aspiring trainers as they are fast and influenced by the most subtle of body language from their human handlers.

Marian, known as 'Mouse' among her friends, died September 25th, 2001 in Hot Springs. I was very fortunate to be able to work with her for a few years. She was a phenomenal teacher and a great person.

For over 25 years, I have had the privilege of working with Bob in many different locations around the world, benefiting from his sharp mind and his wealth of knowledge. Together we have designed training programmes for both general purposes and covert operations. We have trained many animals for operational missions and selected, tested and trained new trainers. Alongside this, we have spent countless hours discussing animal training, while Bob sipped his beloved Drambuie. So the work of Marian, Keller, Bob, and many others, continues and, fortunately, much of it is documented in articles, scholarly papers, interviews and videos. Their legacy also lives on in the many coaches they have trained worldwide, and their influence is evident in my own work.

I have written quite a bit about Marian, but I would like to add how lucky I am to have met her, benefited from her knowledge and wisdom, and her love for animals, training, science and, of course, her family. Her photographic memory was truly incredible. I remember on more than one occasion that she would explain something to me and then ask Bob to bring her a certain book, give it to me and tell me which page to turn to – and she was correct every time! Shortly before her death, I promised Marian that I would do everything in my power to let as many people as possible know what she had discovered about animal training, and I hope I have been true to my word.

Working with Bob Bailey has been a true privilege.

The Baileys' contribution:

The Baileys proved the power of operant conditioning by training a wide variety of animals to carry out complex tasks. Their *IQ Zoo* was used as base for many government-run covert operations, and they were commissioned to train dogs, dolphins, birds, and even cats, to perform intelligence-gathering tasks such as taking photos, and dropping off or picking up electronic devices – all achieved by operant conditioning

DAVID PREMACK

(1925–2015), USA

David Premack did a lot of work involving reinforcers and discovered that responses can be both reinforcers and punishers – it all depends on the animal's state of mind at the time. For example, food can be a great reinforcer. However, if a dog is already very full, food can be more of a punishment; he might prefer a walk. Continuing this line of thought, a hungry dog would prefer food to a walk, and a toy-driven dog might opt to play, at a particular moment, rather than eat. We have all experienced this principle first-hand in some form – the so-called 'Grandma's law' – where, as children, we were only allowed dessert when we had eaten our vegetables.

Premack showed us that a reinforcing behaviour can be used to reinforce a less popular behaviour. For example, I can motivate my son to tidy up his room if he is allowed to play with the playstation afterwards. So, a given activity can sometimes be reinforcing and sometimes punitive in relation to the animal's state of mind, and this can vary with time and place. Experienced trainers, who know their animals well, will take advantage of this principle.

Premack's key findings

- Something can be both reinforcement and punishment, depending on the state of mind, and the situation.
- A reinforcing behaviour can be used to reinforce a less popular one.

The pioneers I have highlighted are our mentors – they have influenced how we structure our training and care for our animals, and some have made us question existing paradigms. I am sure that there will be equally influential people in the future, particularly as science is currently focused on the huge potential in dogs that is still to be realised. It is not only our service and working dogs that are a fantastic source of knowledge and experience; family dogs also have much to offer, allowing us to gain new insights and discover more mysteries of animal behaviour so that we, as trainers, can do better in the future.

GROUND-BREAKING EXPERIMENTS AND THEORIES

In the course of my research for the Pavlov project (*see page 17*), I became aware of significant experiments and studies which challenged certain established beliefs and theories and have, thus, been instrumental in helping me to develop a different view of animal training.

THE ASCH EXPERIMENT

This experiment was carried out several times in different parts of the world, and with different groups of people. It involves a small group of volunteers who are shown a piece of paper with a few lines on it. The lines are numbered. There is a line of a certain length on the right hand-side, and, on the left, there are several lines of differing length. The participants are asked to name the number of the line on the left hand-side of the paper that is the same length as the line on the right hand-side. However, everyone in the group – barring one person – has been asked to name a wrong number – referring to a line that is obviously not as long as the 'pattern line'. Generally, it only takes two or three repetitions before the individual concerned starts to follow the group and names the wrong line. This is because we are social creatures – we like to be part of a group, even if it means we must part company with the truth.

THE MILGRAM EXPERIMENT

This somewhat bizarre experiment also illustrates how easily people can be swayed...

A volunteer enters a small room and is greeted by a scientist in a white lab coat. The setting is important as it creates distance and hierarchy. The volunteer takes a seat at a table, on which there are several buttons and a microphone. The scientist hands over a list of questions and explains that there is a test candidate sitting in the next room (out of sight). The volunteer is told the experiment will focus on the effect of correction. Whenever the test candidate gives an incorrect answer, the volunteer is to press a button on the desk which will give the candidate an electric shock. The importance of the research is emphasised but, even so, it is surprising how many volunteers are willing to continue with the experiment at this point.

So now the test starts and, of course, the candidate gives wrong answers and, therefore, receives electric shocks. But, in fact, it is the volunteer who is the subject of the research; the test is to find out how long they are prepared to continue with the experiment. Every time the test candidate gives an incorrect answer, the intensity of the supposed electric shock is increased. Of course, no electric shocks are administered, but the candidate cries out 'in pain' to make it realistic. Some volunteers stop after the candidate appears to get a really powerful 'electric shock' and screams loudly. But many continue, even when the scientist running the experiment tells them that the current intensity is approaching lethal levels.

We are easily swayed by those in authority telling us what to do. Even when we hesitate, and our inner thoughts tell us to stop doing something we know is 'wrong', peer pressure, or pressure from someone in authority, can make us do it anyway. Once we cross that thin line, it's very difficult to resist and take a different direction. This has particular application in the tough world of dog training – a matter I will discuss in detail in Chapter 17.

THE ZIMBARDO EXPERIMENT

In 1971 an experiment took place at Stanford University's psychology department in the USA, which throws new light on human behaviour.

A 'test prison' was set up in the psychology building, and the participating students were randomly assigned to play prison guards or prison inmates. At the beginning of the experiment, the 'prisoners' were slightly reserved and the 'guards' were slightly assertive. However, it quickly escalated to the point where the inmates were rioting, and the guards were responding with rough, punitive behaviour. The guards began to abuse the inmates, and after a period of six days, one of the inmates started to cry spontaneously and show signs of depression.

At this stage the researchers stopped the experiment because they feared the situation was getting out of control. This is a frightening example of how people think they have to act in certain ways in certain contexts – as do some dog trainers. I have seen many examples of this at dog training venues where force-based training is the norm, showing it is not always the individual who chooses this type of training.

THE PYGMALION EFFECT

The Pygmalion effect, also known as the self-fulfilling prophecy, can cause us problems with how we see the world around us, and the way it affects our behaviour.

Imagine that a coach has been hired to take a group of trainers to a higher level. One of the trainers is experiencing personal problems at home, but he has not shared this with the group. He is struggling to concentrate and has problems understanding the new training concept. The coach begins to think that this trainer is not sufficiently motivated and, at the next session, he spends less time with him. The result is that the trainer's performance deteriorates further. Now the other participants are also beginning to pigeonhole this trainer as 'unmotivated'. A very powerful effect!

This example shows us how easy it is to give someone a negative self-image. But the opposite is also possible: we can just as easily turn the scenario into a positive one. When new teachers were told which children were the brightest in the class, these children performed better than the children who had been labelled as 'bad students'. In fact, unbeknownst to the teacher, the labels 'good student' and 'bad student' had been randomly assigned. When a child, labeled as 'clever', asks a question, the teacher tends to give them more information, more attention, and more time to ask questions.

PARKINSON'S LAW

Another important lesson that deserves our attention is Parkinson's law. It describes typical human behaviour, and was discovered by Cyril Northcote Parkinson in 1958, and described in his book, *The Pursuit of Progress*.

He explains how it is that we use all the available time allotted for a specific task. This is commonly observed in dog training. If a trainer is given six months to train an explosives detection dog, it will take him six months. Even if the dog and trainer are very talented and all the circumstances are ideal, he will use all the available time. I have often checked this against training plans, working out if we could perhaps go through them faster, or with better results. And yes, we could! My first experience of this was with our own laser guide dog training schedule, which was set up in 1996. By then it had taken us nine months to train the dogs so that they could be reliably controlled with laser signals. But after collecting data, analysing the results and changing our training plans, we were able to do it in four weeks - and with even better results.

THE PETER PRINCIPLE

An important lesson can be found in *The Peter Principle*, written by Dr. Laurence Peter and Raymond Hull, in 1969, showing that people always rise to the level of their incompetence. In short, it means that, people are promoted in their careers until they reach a level where they are no longer competent. When this happens, the impact on the person concerned can be devastating; they can lose all their motivation and enthusiasm for their work, and often claim that they no longer like the job in which they were so successful.

Let's take a dishwasher in a large restaurant as an example. He watched the chef prepare the food, and he really liked his job - in fact, it was his dream job. After a few months he was the best and most motivated dishwasher on the team. At this point the chef asked him if he would like to help with the cooking. He was delighted! He learnt to prepare appetisers, then the desserts, then he started to prepare cold dishes and eventually he started to cook. Two years later he was a chef - but he couldn't cope with the stress. He lost all motivation and confidence to the point where he couldn't even wash dishes. He resigned and left the restaurant business. I have often seen talented dog trainers rise to become instructors and lose all motivation in the process - a waste of talent and knowledge.

MURPHY'S LAW

It's also interesting to keep Murphy's law in mind: it helps us think through problems in advance. Generally speaking, it's about things that can go wrong. There are different interpretations of the law such as: *what can go wrong will go wrong* or *things will go wrong in any situation if we give them the chance*. However, dog trainers, in particular, should remember Murphy's law when designing new training plans or beginning practical training.

We need to make sure nothing can go wrong - at least nothing that we can anticipate. When I set up a training programme that requires a feeder, I make sure I have two feeders available - so, if one breaks, I can still train with the spare one. When working with toxic target scents, such as explosives or drug substances, I make sure the dog cannot come into contact with them, but I also make sure I know what to do if this does happen.

Obeying Murphy's Law saved my life when I was employed as a diver, and on many covert police operations. It has also ensured the success of many of our missions because we thought through the smallest details beforehand.

Chapter 2:

WHERE DO WE STAND TODAY?

We have been working with dogs for over 10,000 years – and what have we learnt? That it is difficult! Why? Because we are still not very good at it, and there are a lot of questions to ask. Despite this, there is a seeming reluctance among dog trainers to collect and share information. This could be compared to the memorable moment in history when we 'forgot' how to make concrete – a gap in knowledge that lasted several hundred years, during which time we simply didn't know what mixture to use. It's similar to modern animal training; thousands of years ago the peoples of the Middle East, and the Romans, were experts at training all kinds of animals. Their war dogs were feared on the battlefield, guard dogs protected their homes, hunting dogs were successful in field and forest. In addition, many other animals, such as hunting falcons and eagles, were trained to such a high standard, it ensured their survival. But where has all this experience gone? Where are their training plans? What has happened their knowledge and expertise?

Serious, goal-driven training takes time, money and energy, which is evident if we look at other areas of development. Take the airline industry as an example. In 1903 the Wright brothers flew their plane – The Flyer – a distance of 105 feet – which was sensational news at the time. A few years later, the first military aircraft were used in World War One, initially for reconnaissance and then for dropping bombs. By the late 1950s commercial flights were routine and, fast forwarding another 30 years, fighter jets, such as the F16, were flying at supersonic speeds, remaining airborne for lengthy periods, and carrying highly sophisticated weapon systems. All of this was possible because data was collected and analysed; the engines, the fuel, the propellers, etc., were the subject of constant evaluation and improvement, allowing progress to be made. Learning also came from failed attempts and crashes. Training, schedules and checklists became the norm. Pilots were selected, trained and examined before they were allowed to fly. Engineers had to go through an equally rigorous training programme.

Supremacy in the air was not enough. Mankind wanted to discover more, which ultimately led to the conquest of space. The goal of landing on the moon was spurred on by two significant motivators. One was the race between the US and the Soviet Union to be the first to get a man on the moon, the second was President Kennedy who persuaded his countrymen that this dream could become reality. He not only motivated his people, he also got them to work together to achieve a common, higher goal. A cleaner working in a NASA building was asked about his job. He answered that he cleaned the building, but he also said, with considerable pride, that he was helping to put a man on the moon. It is this spirit which is needed to achieve greatness – working together towards a common goal.

Working with others is often difficult for dog trainers. There is a common saying among trainers: when two dog trainers watch a third at work, they gossip about all the mistakes he makes. What if the Wright brothers had failed to work together, or if the NASA pioneers had laughed at the mistakes of others? None of their ideas would have been realised. Collaboration was the key element that led to success. However, working together means that we must summon our courage, step out of our comfort zone and put egos aside. We need to help each other, instead of judging and condemning each other. We should celebrate success, but also share the pain of failure. These are lessons that dog trainers need to learn. All of humanity's great achievements - all of the seven wonders of the world - are the result of people working together. So, if we really want to learn and achieve great things, we need to learn and be able to work with others. Alone we may make faster progress, but with collaboration we will get further!

CHALLENGING BELIEF SYSTEMS

As dog trainer, the first steps towards changing our thinking, and our ways of behaving, is to examine the current status quo and question its relevance.

There are a lot of paradigms in the dog world. The word 'paradigm' comes from the Greek and describes a model or a pattern. We need paradigms to live our lives and explain what is happening around us. They give us stability and a form of control. When something happens that we cannot explain, we are in crisis. New ideas are tried, followed by a battle between believers in the old paradigms and those who believe in new ones. This is where arguments in the canine world often divide groups rather than bring them together. When I was involved in professional police dog training 30 years ago, one of the paradigms was that you must punish all mistakes, and if you don't punish the animal, he will perceive it as a reward. Positive reinforcement with food or toys was not allowed. A few years ago, the world of police and service dog training experienced a paradigm shift, i.e., there are noticeable changes in accepted paradigms. It is now standard practice in many training groups to use positive reinforcement, and punishment is no longer allowed.

THE IMPORTANCE OF DATA

What we can and should learn from the science of training is to systematically collect data and to evaluate it. So the first question I ask trainers is:"Have you collected data?" Often, they will start defending themselves, saying they didn't have the time or the energy, and explain that they have an excellent memory so don't need data.

Let me be clear: our human minds are incapable of giving us an accurate overview of previous training scenarios. Of course, by and large, we know what happened. But our brains are programmed to give us positive recall. We remember the good events and tend to forget the mistakes, and what went wrong. In many ways, this is beneficial as it stops us becoming depressed. But if we want to learn from our mistakes, we need to collect and analyse data. Don't get me wrong, it's not that our brains are unable spot mistakes. If we didn't learn from our mistakes, none of us would reach old age. However optimum learning involves collecting data because it provides objective information, giving us the means to make our training faster, more reliable, stronger, more efficient and more effective.

My own training has become much faster and more reliable through data collection, particularly in the case of training camera dogs. When I started working with police dogs, we 'fired' police dogs if they were unsuccessful – no one else was to blame. Nowadays we also fire trainers when they fail to make the grade. This can be a sore spot among training groups who are struggling with data collection – and it is mostly ego and fear that get in the way. Often it is the group leader who fails to convince their trainers of the importance of data collection, and this is often a sign that a safe, secure work environment does not exist within the unit.

Why should a good trainer have problems with the documentation? It shows us how we can change behaviour and how quickly we can produce well-trained dogs, which is ultimately the job of a professional dog trainer. Good trainers take pride in what they do. Their skills, experience and data contribute to better, stronger and more reliable behaviours in the dogs in their care. The dogs increase in value and usefulness, as do their trainers.

Quality is key, so we ask: how good is good enough? How do we measure what is 'good enough'? What can we expect from a well-trained dog and handler? What can we expect from an experienced trainer? Do we have to waste another 10,000 years wondering why dog training is progressing so slowly? Or do we finally start collecting and evaluating data and sharing the information with others?

UNDERSTANDING GREY AREAS

When I am working with people in managerial roles, I point out that there are a lot of similarities between animal training and dealing with employees (see *page 250*). Managers learn a lot about different approaches because no two people are the same. Individuals have different motivations, and, equally, animal training is not black or white. Dogs differ not only in their breed, but also in their personalities – just like dog trainers...

Besides differences in personality, there are other factors that contribute to the fact that training can never be just black, or white. We communicate and learn in different ways. We have different experiences and different prior knowledge. We differ in our powers of observation, in our sensitivity, and in many other things. Sometimes we wish training could just be black or white – because then it would be easy. We could follow a clearly-defined black and white plan or protocol and everyone could train and work with dogs to a defined level. We would achieve the same uniform results, and training would always take the same amount of time.

But in animal training, as with human training, there is a grey area, and there are no textbook instructions for dealing with it. Ideally, there would be no difference between the theoretical and the practical, but this is rarely the case. There is a phase of training where people listen to their gut instincts. It is the moment when we start to train not only with our brains but also with our hearts. Emotions come into play; we start looking for different points of view and instigate discussions. We put the training in context. Distractions come into play, the trainee's motivation changes, and the mode of learning lacks clarity. It is the moment when our cues or signals no longer lead to predicted behaviour because the trainee's 'black box' (see *page 124*) can change the outcome in unexpected ways. It is an exciting time for experienced trainers, but a risky period for novices, and many throw in the towel, deciding that training is not for them.

When discussing the grey area of training, a colleague painted the perfect picture:"to be able to see grey, we must first be able to see black and white!"

Chapter 3:

DOGS ARE AMAZING!

So far, we have looked at training from the human perspective, celebrating the work of pioneers, and examining the flaws in our established ways of thinking and working. Now it's time to look at training from the dog's perspective. What is our trainee thinking and feeling, how does his brain work, how intelligent is he, and how can we optimise learning?

Dogs are truly fascinating creatures. They are closer to us than any other animal; they have become attuned to our ways and show us real attachment. They can read our body language, interpret our looks and smell whether we are happy, stressed or scared. They can even detect when we are ill, evidenced by medical detection dogs. Dogs are our helpmates; on the hunting field, guarding and protecting our property, using their incredible sense of smell to find missing persons, as well working alongside the police and military in a variety of roles. But above all, they give us love and joy, enriching our lives and contributing to our mental and physical wellbeing.

A relationship with dogs requires us to connect with another world, namely that of animals. These days, we are so taken up with technology, we are in danger of forgetting how to communicate with other people. A phone call is rare; we prefer to make and maintain contacts via email, Instagram, Facebook, Twitter and many other apps. But having a verbal conversation with someone, where we respond to eye contact and read body language, is something we rarely experience. And before we know it, these social skills will be lost!

Dogs do not use high-tech devices to interact with us, and their vocalisations are not sufficient to tell us what they need and how they are feeling. We get most of our information from observing their behaviour. But if we are struggling to relate to fellow humans in this way, how good are we at observing our dogs and analysing their body language?

Among English businessmen there is a saying: *no contract without contact*. Despite our reliance on technology, when it comes to the crunch, people would rather meet 'in person' rather than closing the deal via email or a zoom session. In dog training circles, this means that if we want to make a 'contract' with a dog, we need to be in real, close contact – enjoying long walks in the woods, going out and about, or simply sharing companionable time together at home. Dogs are social creatures and need access to social networks – real ones, not virtual ones. They do this in a variety of ways. They enjoy physical contact with other dogs in the park. They sniff each other's butts, send messages by urine-marking, and read each other's body language. They communicate their feelings to other dogs using a large repertoire of body language, and they try very hard to understand our human body language as well. Of course, a dog does not need to be friends with every dog he meets. Like us, dogs have their preferences, seeking contact with those they like, and ignoring or avoiding others. Dogs are not loners! Their social network is important; they need contact with others.

When Marc Bekoff wrote *The Emotional Lives of Animals* in 2008, the world changed for many trainers. Bekoff presents dogs as emotional, sentient beings, and this was a game changer for many. Dogs are good at comforting us when we are lonely, sad or anxious. We don't have to teach them or ask them to do this, nor do we have to reward them with food or toys for their attention. They just seem to know when to come and make contact. Dogs give us unconditional love and friendship – there is no hidden plan, no bad intentions – what we see and get is clear, pure and honest. No wonder we like working with them!

We should also consider dogs as individuals, just as we would when working with human colleagues. We label people as extrovert or introvert, calm, confident or fearful. There are sensitive or responsible types, and many variations in between. We appreciate the importance of understanding the individual in order to find the best approach when working with them. It is just the same with dogs. Canine personality types are the subject of some debate, with talk of reactivity, sociability, subservience, aggression, anxiety, willingness to work with people, and much more. It is not my intention to discuss personality types here but, rather, to draw attention to the importance of treating every dog as an individual and selecting a training method that is suitable to their particular needs.

ASSESSING INTELLIGENCE

For a long time, people have been interested in intelligence quotient (IQ), but recent studies have found that emotional intelligence, EQ, is also very important to our wellbeing, success, and even our survival. And yes, there are differences between individuals; we are not all the same, and neither are our dogs. There is a huge difference in their intelligence, and this has nothing to do with body size, skull size, or brain size. Just look at the incredible intelligence of some birds, such as crows; their brains are much smaller than the smallest canine brain and yet they can complete the most complicated tasks. I have done a lot of work with chickens to learn more about operant conditioning, and I was amazed at what their little brains could do! I have seen birds do things that dogs cannot do – and this has nothing to do with physical capabilities but, rather, how they use their brains, their memory, and their problem-solving skills.

The American psychologist and dog trainer, Stanley Coren, studied the intelligence of over 133 different dog breeds and found that their communication is more varied than we thought. The way they think, and how they use their memory, is more sophisticated than we imagined and is heavily influenced by character, inherited traits and diet. Coren compiled a list of the 79 smartest dog breeds, but it is important to understand that this is not just about intelligence and breeds, it also involves personalities and life experiences.

But what is intelligence? It is not so easy to define. Does it relate to our recall of previous learning experiences? Or is it measured by how well we can solve problems? Hounds are very good at tracking, but don't ask them to herd a flock of sheep. Border Collies excel at controlling sheep, but they don't make the best protection dogs for the military. At this point, scientists are still not sure how to measure a dog's intelligence. What we know from our own experience. is that they are incredibly good at listening to and understanding our human language.

Our radio-controlled special operation dogs are sent far out of sight of their handlers, often in the middle of the night. Sometimes they are sent into buildings where they can no longer see their handlers. Their directional cues are transmitted over the radio: *"go left"*, *"go right"*, *"forward"*, *"come back"*. They are working in a completely new environment, while completing complex tasks, such as looking at a specific item (for the camera to record), jumping over, or crawling under obstacles, carrying items to place in a specific location, or finding items and then transferring them to a new location. All the while, they are dealing with any number of distractions. These dogs exhibit a series of behaviours, over a long period of time, without seeing a human and without receiving a reward. They are working solely in response to commands received through their headsets. When the mission is completed, we are always amazed at how easily the dogs find their way back to us when they get their *"come back"* signal. It's as if they have stored a mental map of the paths they have travelled.

Psychologist Nicole Chaphuis tells us that dogs are capable of creating mental maps of an area. In field trials, she found that dogs appear to identify landmarks on a route and carry a mental picture of their own position in relation to these 'signposts'. To compare this with human capabilities, I can confirm from my operational experience that if I sent a remote-controlled robot into a building to 'inspect' a series of different rooms over a period 20 minutes, I would find it extremely difficult, if not impossible, to direct the robot back to the start position.

The camera dogs we train for covert operations are now remotely controlled via a modern headset.

PROMOTING COGNITIVE ABILITIES

The intelligence or cognition of dogs is an exciting topic and, as trainers, we can learn a lot from the ongoing research in this area. When working on specific tasks, this knowledge can help us select the most suitable trainees and, going forward, it may change the way we train our animals. We may be able to ask for behaviours that we cannot even imagine right now.

We know that children with a very high IQ are easily bored at school, which is why it is important to identify them and to support them appropriately. The same goes for dogs. When we understand more about their intelligence, we can design specific, empowering training programmes for them. Research is ongoing. Brian Hare at Duke University, North Carolina, USA, is currently studying the evolution of cognition in humans, the great apes (our closest relatives), and in dogs. Other interesting research comes from Gregory S. Berns, an American neuroscientist, who has used a magnetic resonance imaging (MRI) scanner to see what is going on in a dog's brain. In particular, he examined and recorded the activity of the area in the brain that is activated when a dog is feeling happy. First, he observed dogs, via MRI, while they were given delicious food, such as a sausage. He then measured their brain activity when they could see, or smell their guardians, or received cues from them about upcoming rewards. The brain activity he observed gave concrete proof that the actions of us humans have a significant impact on the lives of our dogs.

A number of games have been designed to test a dog's cognitive abilities, and these give us the opportunity select the individuals we need for specific programmes. Of course, a scent detection dog needs to use its nose more than its eyes – but it's not just about the sense of smell. These dogs also need to work with their handlers and solve problems. For example, how does a sniffer dog deal with a target scent that is high up and out of reach? How does he indicate a find when he cannot get close to it? Our 'remote controlled' camera dogs are tested on how quickly they can solve problems. Can they generalise from what they learnt in training to an operational scenario, and how quickly can they do this? Can they relate their actions to outcomes or consequence, at speed? How quickly do they learn complicated tasks and chains of behaviour?

Regardless of what we want from our dogs, we need to make sure we keep them busy physically, but especially mentally. Boredom is the enemy. Dogs thrive on being given a challenge, and enrichment from their environment and in their everyday interactions, is key to their wellbeing.

THE ROLE OF MEMORY

Dogs are extremely good at remembering things – just look at how they interact with familiar people, environments, buildings, smells and sounds. We tested this on a number of young scent detection dogs, who had to quit the training programme due to health problems. These dogs, who were part way through their training, were rehomed as companion dogs. Some six years later we checked to see what they had retained from their previous training. In all cases, they still recognised and correctly indicated a find when presented with the odour of drugs or explosives. However, a recollection, stored in the memory, is highly unlikely to become active unless there is a trigger.

We believe that it is only us humans that have the ability to think about the future: what we want to do, how we might react, and the potential consequences. It is thought that dogs are incapable of doing this and, therefore, they cannot 'plan' an action or a strategy. Imagine you and your dog are walking in a park. As you cross a narrow bridge, a mountain biker comes out of nowhere. He approaches at speed, and then suddenly jams on his brakes. It is only a quick side-step, on your part, and a tug on the lead, that prevents a collision. Two weeks' later, you relate this incident to a friend in your living room. You remember everything: the noise, the shock, and your dog's reaction. You can remember exactly what happened and replay the scene in detail. Your dog, sleeping next to you on the sofa, probably doesn't remember the incident at all. You cannot get him to relive the scene; he may not even remember the scary experience when you go back to the bridge in the park. However, if a mountain biker suddenly appears, makes a noise, and your dog is on a leash, the chances are his memory will kick in.

This insight into how memory works in dogs tells us a lot about how we need to design our training programmes. We need to build in clear triggers that initiate behaviour when we need it. We must also ensure that these triggers are not linked to other signals, such as the (training) environment, other noises, smells, etc. For example, if we only train a scent detection dog within the confines of the training ground, he will never find the target scent outside this environment.

MIND READERS?

If you have worked with dogs for a long time, you will probably have experienced those moments when it feels like your dog can read your mind. In our operations we had these moments again and again, especially in situations where we were under a lot of pressure.

Intelligence is to do with making connections and understanding that good or bad consequences will follow. We are convinced that trial and error is not the only factor responsible for learning in dogs; there has to be more. In operational scenarios, we have seen dogs show behaviours that have not been taught and could not be taught to other dogs in similar situations. But in life-threatening scenarios, dogs have shown they can act on their own initiative and, as a consequence, they have saved lives.

The story of Clever Hans – the horse who could, supposedly do arithmetic, but was, in fact, following cues from his trainer (see *page 141*) – has taught us that the most subtle of signals can influence behaviour. There have been many instances when scientists have tried to validate reports of 'mind-reading dogs'. However, the moment they eliminated subtle signals given by guardians, including environmental cues, the dogs could no longer 'read minds'.

When evaluating such reports, we also have to bear in mind that humans have a selective memory. The scientists therefore made sure that they worked with objective observers and video recordings so as to get accurate information. For example, the dog who gets up in advance to greet his guardian, also gets up when other visitors or the postman, approach the front door. However, this may go unobserved and all we remember is the dog's behaviour when his guardian comes home. We then begin to construct a story and believe that the dog can predict when his guardian is coming home. This tells us more about human psychology than how the dog's mind works!

Don't forget that dogs are born with a wide variety of innate behaviours and abilities which sets them apart from other animals. They enjoy being around people, and other dogs, and they recognise dogs from their own group. They are attracted to things that move, and they perceive the world largely through their sense of smell, which can trigger actions that we cannot understand.

KNOW YOUR DOG

We have already discussed individual differences between dogs, such as personality traits, breed dispositions, and innate behaviours, and how this knowledge should be applied to designing training programmes. This is more than straightforward relationship building, it is about understanding how a dog sees the world. Do we know:

- What makes him happy?
- What scares him?
- What challenges we can give him?
- What distracts him?
- How he deals with stress?
- What calms him?

We need to observe our dogs. The best place to start is within the home. We need to observe what we see – and write it down. Start collecting data!

After making initial observations at home, we can progress to looking at behaviour on walks and in other environments:

- Does the dog stay close, or does he want to explore?
- Does he use all his senses – ears, eyes and nose – or is one of them dominant?
- How does he approach other dogs and other animals?
- How does he behave at the vet?
- How does he react to walking along a busy street or travelling on public transport?

Before we start training, we need to gather all this information, and to spend both time and energy building a good relationship.

Get to know your dog and discover everything to do with his individual nature.

Chapter 4:
SENSORY ABILITIES

Before starting an expensive and complex training programme with dogs selected for operational duties, we first give them a thorough physical check-up. We X-ray hips, elbows and spine, we check blood, heart function, overall condition and much more. We also look at a dog's sensory abilities, because any issue with sight, hearing or sense of smell will not only impact his working role, it will also affect his training, risking lengthy delays. There are some simple tests that can be carried out, but specialist vets, with the latest equipment, are required for an accurate, and objective examination.

SIGHT

We humans are visually oriented, whereas as our dogs are more concerned with their sense of smell; they, therefore, perceive the world in a different way. That is not to say that dogs don't use their eyes - of course they do - especially when objects are more than half a metre away. At close range, they do not see clearly and are more likely to rely on sniffing or touching objects to gain information.

Like humans, there are dogs who have problems with their sight and, in most cases, this would rule out active duty in the services, and in sports such as agility. We also need to understand how a dog sees, as the visual acuity of humans and dogs is different. We humans can see an object at a distance of around 20m, whereas a dog would have to be at around 6 m from the object to sight it clearly. A dog's vision is, therefore, inferior in terms of clarity, or 'sharpness', at a distance.

There is also in difference in our perception of colour as we humans see the world in a much wider spectrum of colours. Dogs are not colour blind, but they are red-green blind, meaning these two colours look almost the same to them, and are perceived as grey-green. However, they are good at distinguishing between yellow and blue. Colour vision becomes important when selecting targets or reward toys for training. A ball that we see as bright red is virtually indistinguishable to a dog who is running after it on green grass. We took this into consideration when we started using laser pointers for directional training and ruled out red targets. We also worked out a dog perceives a laser when it is aimed at a specific background, which may reflect or absorb the light, or even change the colour spectrum.

Besides colour vision, we also need to look at field of vision (angle) and sharpness/clarity. In general, we humans - unless we need spectacles - see clearly most of the time. In contrast, a dog's vision is somewhat blurred. It is important to keep this mind when working with targets, hurdles, and other sight-oriented tasks. When our camera dogs returned from a mission, they often had a hard time finding their handler among all the team members, dressed in their bulletproof protective clothing. To solve this problem, the handler dispensed with their protective gloves so their bare hands were a point of orientation.

Colour vision human

This is what we see.

Colour vision dog

The dog's perception is approximately that of a red-green blind human.

Humans have a larger area of sharp focus (green) than dogs, but the dog has a wider range of vision to the sides and behind (grey), although it is less sharp in these areas.

A dog's field of vision is much wider than that of a human. A dog sees much further to either side than we do because their eyes are located to the side of their heads. However, the area in which the field of vision of both eyes overlaps is smaller in dogs. This affects the depth of field; a dog can only see something clearly when it is directly in front of him.

There are also breed specific differences caused by conformation. For example, a retriever with a broad muzzle has a wider field of vision than some smaller and longer-nosed dogs. This needs to be taken into consideration in training situations. Initially, a dog with a narrow range of sharp vision should be presented with his target frontally and not to the side. In training, we should also consider the dog's wide field of vision which means he perceives more distractions 'out of the corner of his eye' than we do. We also know from research that many gundog breeds are far-sighted and many typical sheepdogs are near-sighted.

Dogs can see better at dusk than humans; they move quickly and safely in low light situations where humans would struggle. However, they cannot see in total darkness. When we tested dogs in pitch-black bunkers or basements, they began to move in a similar way to us humans: they would slow down, run into objects and there was a point when they would no longer dare to move. We solved this problem by attaching a small lamp to their collar or camera helmet.

HEARING

In addition to their sense of smell, a dog's hearing is also exceptionally good. Young children can hear sounds up to 20,000 hertz. As we age, these levels decrease, and by middle age we can expect to hear up to around 14,000 hertz. According to research, dogs can hear sounds up to a frequency of 45,000 or even 65,000 hertz, which is incredibly high. This is thought to be related to their ability to catch rodents, such as rats. A rat's warning sounds have a frequency of 22,000 to 44,000 hertz, which is inaudible to us humans, but can be heard and located by dogs.

There are many other instances of dogs using their hearing in addition to their sense of smell to perform a designated role. Hunting dogs use their hearing to find wounded animals, search and rescue dogs use their ears as well as their noses when finding victims of earthquakes, and police dogs make use of their hearing as well as their sense of smell when tracking down suspects. During the Vietnam War, the US military found that dogs detected trip wire mines not only by smell, but also because they could hear the ultrasonic noise (over 20,000 hertz) made by the wind blowing over the trip wires.

Many family dogs will jump up off the sofa and run to the front door when their guardian comes home, long before the other family members have heard them coming. They can identify a particular car engine if it has significance for them and detect whether familiar or unfamiliar footsteps are approaching the house.

In all, a dog's sense of hearing is much finer than our own. Compared with us, a dog can hear in a wider range, from a greater distance, and is better at discriminating between sounds.

SMELL

What most impresses scientists and dog handlers alike is the dog's nose. Over the years, the dog's sense of smell has been the subject of extensive research and, more recently, technology, such as MRI scanning, has been used to see what is happening in the dog's brain when he smells familiar caregivers, his favourite treats, or target scents such as drugs or explosives. All of this helps us to create better training plans. However, we still have much to learn, particularly concerning the limit of a dog's scenting powers. This is an animal whose sense of smell is not only vastly superior to our own, it also outranks many sophisticated, technological devices.

A good example is the work of an arson dog. If arson is suspected, experts use hand-held devices to look for traces of fire accelerants. The drawback is that they only use their measuring device in those places which, they think, could be the source of the fire. In contrast, an arson dog searches in places that people would never have thought of. He just goes where the smell molecules take him. It is often the case that the hand-held measuring devices fail to give a reading at the points where the dog is indicating! When the experts take samples from the scene of the fire and examine them in the laboratory using gas chromatography coupled to mass spectrometry (GC/MS), it has been found that the dog's scenting ability exceeds the measurement threshold of the GC/MS. This is absolutely amazing! It is also one of the reasons why arson dogs are trained with odour concentrations measurable by GC/MS because this concrete evidence is required when reporting from a crime scene; a dog's indication, on its own, is not enough.

CONSCIOUS PERCEPTION

Dogs are very aware of their surroundings. Have you ever seen a dog step into another dog's poop? Very rarely, if ever. This is because dogs are very conscious of the environment and where they put their four paws. Traveling in a car to a favourite walk, a dog may become super alert on the last stage of the journey as he senses his whereabouts. He may have been asleep for hours, but suddenly awakens when he is close to home after a trip away, or as you approach the car park of the training venue. Dogs appear to create a mental map of their environment, and they know exactly where they are within it.

Our senses absorb a lot of information, but it only comes together in the brain. Only here can we process the information conveyed by our senses and thereby understand the world around us. As we have seen, humans and dogs do not perceive the world in the same way as our senses are feeding us differing information. However, we humans tend to believe that our perception of the world *is* reality, while, in fact, our dogs live in a very different reality.

Imagine being able to see the world from a dog's point of view. You lie on the ground and look down the street. Everything is blurry and it is very noisy. The colours are yellow, blue and grey-green. You smell olfactory messages everywhere and you know exactly where you are all the time. If we understand that dogs do not live in the same world as us, we should not expect them to behave like us either. They will not make the same decisions as we would, and they will deal with challenges differently; they absorb more information and process it in a completely different way.

Chapter 5:

GETTING DOWN TO BASICS

Human psychology deals with the human mind and its functions; the main goals are to describe, explain, predict and change behaviour. This applies equally to the training of animals - it all starts with observing behaviour. Therefore, before we start working out how to change behaviour, we must first define what it is.

Quite simply, behaviour is everything that an organism - human or animal - does; movement, actions, and reflexes. Behaviour takes many forms, ranging from isolated actions in which only a few muscles respond to actions that involve the whole body. We apply the term 'reflex behaviour' to some of very simple and primitive innate physiological responses, but this can also apply to more complex behaviours, such as nest building in birds, pack hunting among wild dogs, or a detection dog searching for his target scent. Behaviour in both humans and animals is continuous; it occurs all the time and, in all situations - even when we cannot see it.

There is a simple way to define behaviour. Animal trainer Bob Bailey calls it 'the dead man test':

Can a dead man do it?
If so, it's not behaviour.
Can a dead man lie still and remain motionless?
Yes, so lying still is not a behaviour.
Can a dead man ignore sounds?
Yes, so ignoring sounds is not a behaviour
Can a dead man operate a clicker?
No, so using a clicker is a behaviour.
Can a dead man give a dog reinforcement?
No, so giving reinforcement is a behaviour.

The dead man test:
Can a dead man do it? If so, it is not behaviour.

We can easily convert the 'dead man test' into a 'dead dog test' to better understand the concept of behaviour:

Examples of dog behaviour includes: running, jumping, catching a ball, searching, attacking, barking, eating and biting.

Examples of non-behaviour include: not getting out of the car, not chasing a ball, not eating, lying still.

So, if you are wondering if something is behaviour, ask yourself if a dead dog can do it. Of course, there is a catch. Lying still is something a dead dog can do. But when a living dog lies still, behaviour is still occurring. Reflexes make the dog breathe, his digestion is still active, and his heart is beating. Sometimes the 'dead dog test' is a little too simplistic, but it gives us a focus.

OBSERVING, ANALYSING AND CHANGING BEHAVIOUR

If we want to change a behaviour, we have to observe it, collect data and then analyse the data. We need to understand what motivates the behaviour, identify the triggers, and then consider how to bring about change.

Let's take the behaviour required for a dog to negotiate an A-frame on an agility course as an example. The dog must approach the A-frame at speed, run on to it, and over it and, as he completes the obstacle, he must hit the contact area. After a few repetitions, the dog will pick up speed and will be tempted to jump from the down-side of the A-frame, incurring penalities, rather than running through and hitting the contact area.

This is where the use of slow-motion video footage comes into its own – it is the perfect tool for observing and analysing behaviour. When we replay the footage we can see exactly what the dog is doing: how he moves and the point at which he jumps off the A-frame. We can question why he is moving so fast: what is triggering his behaviour, and what is acting as reinforcement? If we are able to answer all of these questions, we are in a position to write a training plan detailing how to change the dog's unwanted behaviour to wanted behaviour.

Behaviour is changed by its consequences. It's simple, but not easy. In general, we humans seek wellbeing and try to avoid discomfort or pain. There are reasons or motivators that change our behaviour. The same applies to animals. Changing a behaviour, or getting a new behaviour, is not difficult if we break it down into small parts. We all learnt to walk; we started by crawling, then taking the first small steps, falling down, getting up again and eventually walking – then progressing to running and jumping. Skills can be developed further, so we have athletes who can run faster and jump higher than the rest of us. The process of learning involves a series of steps that form a chain and, at the end of the chain, is the finished behaviour. If we can see behaviour in this this way, we will be able change it, and/or introduce new behaviours. Humans and animals learn 24 hours a day, seven days a week. We cannot turn off learning; it's happening all the time.

Imagine a puppy using his nose for the first time, trying to walk, run, and play. In the process, he is developing his muscles, finding his balance, and learning new skills, such as moving at speed, negotiating obstacles, and holding things in his mouth. The successful consequences of his actions result in the acquisition of new skills. His muscles get stronger, his senses sharpen and his nervous system is receptive. Many thousands of reactions influence his movement and his actions; interactions with the environment are reinforced, weakened, shaped or differentiated, ultimately providing the tools for survival and reproduction. They are the result of successful or unsuccessful actions, and the behavioural changes they bring about are what we define as learning. Remember, all living things, including humans, learn as long as they are alive. Learning is not confined to classrooms, reading books or taking online courses.

People and animals learn 24 hours a day, seven days a week.

Many trainers think that dogs switch into learning mode when they enter the training venue and will, without guidance, continue to follow the rules in all other situations. Learning never stops; the issue is that it might not be the learning we want or consider to be desirable. Therefore, the dog who can do immaculate heelwork at the training venue, may learn to pull on his lead as he heads back to the car. It's the same with young children. At school they learn to behave and stick to rules – *we mustn't fight, we mustn't yell at each other, and so on* – but at home or when playing with friends, they may yell and fight and engage in other undesirable behaviours. Children learn around the clock, too!

Learning has to start somewhere, and it makes sense to choose a safe environment without too many distractions. But, as soon as possible, we need to go out into the 'real' world. Most behavioural changes in animals occur during interactions with the environment, therefore, we, as trainers, need to exercise some control over this to encourage effective learning. We can do this in small steps, gradually introducing more and more environmental distractions, until the dog can not only cope, but carry out complex operational tasks, no matter what is happening.

Animals solve problems to get what they want. This can be food, water, sex, shelter, attention, or whatever is currently of value. If we present a dog with a simple problem in a safe, distraction-free environment, he will solve it to get the reinforcer he wants. If these tasks are gradually made more difficult, he will become a skilled problem-solver, helping us accomplish tasks that would be near impossible without his input. For example, we struggle to find drugs that smugglers have hidden in cars, planes, or boats. Dogs, with their unique sense of smell, can help. So, we start training in a safe environment with no distractions. If we set the dog the task of finding his trained target scent i.e., the scent of the drugs, from a line-up of 'suspects', or a scent carousel – a training device which presents target and non-target scents to trainees – he quickly learns that there is a reward for indicating a 'find'. Gradually, we add more and more distractions, not just in the line-up or the scent carousel, but also within the training environment. Eventually we step out of the room into real life, and prepare the dog to find drugs almost anywhere, even in seemingly impossible situations such as in a crowded airport, a swaying ship, or in a cramped car. We teach the dog that there can be small or large amounts of his target scent which may be easily accessible, or well hidden. In addition, he may be working in an environment imbued with distracting smells – other animals, for example – or the scent of drugs may be masked by a particularly pungent odour, such as coffee. We practise in light and dark environments, in hot and cold temperatures. At every step, we 'pay' the dog with reinforcers. The whole process is broken down into small steps, which entails working with training plans and protocols. This can be likened to school timetables or to construction plans for a house.

"WHAT'S IN IT FOR ME?"

If a dog doesn't gain anything from performing a 'desired' behaviour, he is unlikely to repeat it. Why should he retrieve a dummy, search for narcotics, or complete an agility course?

Remember Thorndike's law of effect: *Behaviour will be changed by its consequences.*

We should also consider behaviours that are influenced by environmental consequences; they persist or are lost because their consequences have resulted in strengthening or weakening. We call this process 'learning' and it brings about, more or less, permanent changes in behaviour through environmental experiences. Bear in mind that all behaviour, including learnt behaviour, is based on natural behaviour that has been developed or modified. We cannot train a behaviour that is unnatural to the animal. This is illustrated by the way our children's development is governed by learning through trial and error. Every moment of the day, a growing child becomes increasingly proficient as they develop the skills of co-ordination – walking, running, climbing, holding a cup, etc. – building on success, and learning from failure. In just the same way, an animal trainer modifies, and develops, existing behaviours to reach the intended goal. There are some taught behaviours which appear to be unnatural, such as a dog walking on his hind legs and other choreographed moves he may be asked to perform in a canine freestyle routine. However, this is not the case: these moves are essentially natural behaviours, otherwise the dog would not have been able to learn them.

We cannot train a behaviour that is completely unnatural to an animal. We can only modify existing behaviour.

WANTED OR UNWANTED BEHAVIOUR?

Before we continue, it's important to understand the difference between wanted and unwanted behaviour. Both differ from individual to individual and are influenced by time, context, place, circumstances, emotions, experiences, and many other elements.

Let's look at the example of a dog barking indoors. It can drive neighbours crazy, especially if it occurs every time the dog is left on his own and continues for a lengthy period. Clearly, this would be considered unwanted behaviour by just about anyone. But if that same dog suddenly barks in the middle of the night to warn his sleeping guardian that burglars are tampering with the door, that would be wanted behaviour. And if the same dog chased a thief who was about to steal a car, then all the residents in the street would be very happy with him. This shows that wanted and unwanted behaviour is not always easy to define.

What about a police dog? When out on a night patrol, he should bite the suspect who gets too close to his handler. But, of course, he should not bite the neighbour who approaches for a chat. We determine what is okay and what is not; what is wanted or unwanted. However, if we are not clear about what is desirable and what is undesirable, it will be unclear to our dogs. This issue presents one of our biggest training challenges; we therefore need to create training plans that provide clarity for the trainer, the handler, and the dog! We need to work with solid criteria and to be objective so that we can collect the necessary data during training that will allow us to measure success.

I like to hand my training plan over to a non-specialist, who is not involved in training animals, to see if they can identify the desired behaviour I am seeking to teach. Alternatively, I might show a video recording of a training session to see if the goals I am working towards are clear. If a layperson struggles to follow my plan, how can I expect my dog to understand what I want?

When working on a training plan we should also bear in mind the following points:

- Some behaviours are easier to train than others.
- If a desired behaviour is very weak, or occurs infrequently, it is difficult to reinforce.
- A behaviour may also be blocked or overshadowed by a competing behaviour.
- Some behaviours are incompatible. For example, a dog cannot bark and hold something in his mouth at the same time.

We need to appreciate that an animal – in this case, a dog – has a wide range of behaviours which involve much more than meeting the needs necessary for survival.

A training plan involves giving the dog a clear understanding of the intended task.

Animal trainer Bob Bailey is a great believer in: *"You get what you reinforce"*. Quite simply, it means you get behaviour that is reinforced, motivates and is, in some way, worthwhile for the dog. This fits perfectly with the *"what's in it for me?"* attitude, already discussed (see *page 49*). If a particular behaviour produces something the dog likes or wants, he will keep repeating it, and it will increase in intensity.

"You get what you reinforce."
Bob Bailey

For example, what does a dog get from pulling on the leash? A typical scenario is for a dog to drag his guardian from tree to tree and from lamppost to lamppost, making the walk a misery. In fact, pulling on the leash is one of the most common training problems experienced by guardians of pet dogs. Once we consider it from the dog's point of view, it's very easy to see the reason for this behaviour, and why it is so hard to stop. Dogs love getting information from other dogs – and passing on information about themselves. It is exactly the same as us humans sending and receiving messages on our smartphones. Trees, lampposts, and other 'landmarks' contain a lot of information in the form of urine, which I refer to as 'pee-mails'. It's like a newspaper or a big sign that says: *Hello there! Read this and tell me what you think about it!* The dog's nose receives the information from afar: *New message!* His guardian is unaware of this important message so when they turn in a different direction, the dog pulls back towards the scent. Now he can read the new message and leave his reply. This is very motivating for the dog; the answer to *"what's in it for me?"* is very clear. His choice to pull on the leash has proved to be a successful strategy, and, in no time, it becomes an engrained behaviour. As far as the dog's guardian is concerned, this is undesirable behaviour, but for the dog, it is positive, desirable, and empowering.

If we understand the difference between wanted and unwanted behaviour, we can change it. To do this, we need complete clarity with regard to our goals before we start working with our dogs. This enables us to 'design' a behaviour, thereby making the change in behaviour easier for all those concerned - guardian/trainer and dog. Clarity replaces doubt, hesitation, confusion, and frustration, putting those involved - at both ends of the leash - on a path where they are more likely to succeed than fail.

THE DOG'S PERSPECTIVE

Dogs and humans have a different way of thinking. Dogs live in the now, and do not consider the past or the future. We humans label behaviour as wanted or unwanted - dogs do not share that viewpoint. They react, and their behaviour focuses on what they want or need. If they want to pee on a tree, they will, if necessary, drag their guardian to the desired spot. A dog who growls when someone approaches him while he is eating is making sure that nobody steals his food. The dog who barks when left home alone is trying to get his guardian to come back. And the dog who bites when a stranger enters his house is trying to defend his territory. All these, and other behaviours, are motivated by the dog's desire to stay in his familiar and loved comfort zone, to deal with unfamiliar situations, to reduce stress and to combat anxiety.

Dogs often join up the 'wrong' dots and continue to show unwanted behaviour because the consequences are worthwhile for them in some way. Barking at a stranger when out on a walk causes the stranger to walk away. Rummaging in the rubbish may lead to finding a piece of meat. Breaking these connections can be challenging, especially when aggression is involved, and the boundaries are unclear. When dogs are just 'loved' by their caregivers, and don't understand the meaning of "*no*", the consequences are often disastrous. That well-known saying - *love will be the death of me* - can be applied to dogs who exhibit excessive levels of unwanted behaviour or aggression. Sadly, they end up in animal shelters and may be euthanised, purely as a result of poor training and poor guardianship.

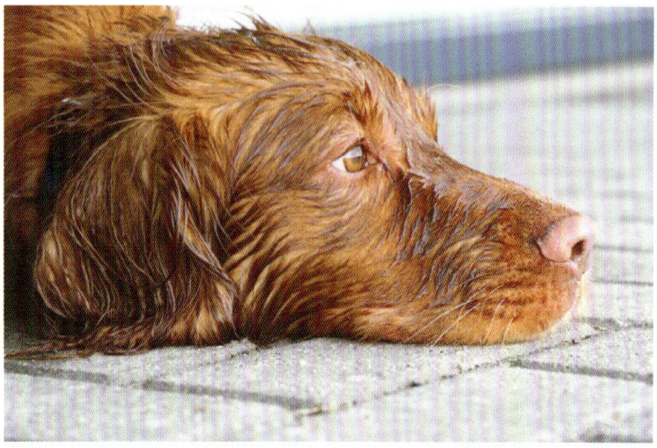

There are times when a relationship needs to be reconfigured, allowing confidence to grow.

Ultimately, it is our responsibility to help dogs understand what behaviour is acceptable in our society, and what is not. Setting boundaries is important and ensures everyone is on the same page. I have seen cases of dogs terrorising their household - and the entire neighbourhood - because their problematic behaviour had escalated to the point where they were out of control. Obviously, this poses a threat to everyone who is involved - including the dog. As behaviour experts, our aim is for everyone to enjoy a good quality of life - people and dogs alike. When we have the opportunity to re-set the relationship between a dog and his guardian - setting new boundaries and creating new connections - we can improve both safety and comfort, and confidence will grow.

For more information, see working with negative and positive consequences, page 59.

Chapter 6:
THE POWER OF CONDITIONING

The impressive work of the early of pioneers of animal behaviour has given us many insights into efficient and effective training methods, and some of their principles, discovered over 100 years ago, are still in use today. As already highlighted, their experiments were designed to learn more about human behaviour, and the basic finding – that behaviour changes because of its consequences – holds true for humans and animals alike:

- When it starts to rain, people, like animals, seek shelter.
- When the fire becomes too hot, all living beings seek distance from it.
- When hungry, both animals and people look for food.
- This gives us an understanding of motivation – the reason why animals and humans behave, or don't behave in a particular way.

Behaviour changes as a result of its consequences.

The benefit of scientific research is that we can study it and learn from the principles that emerge. The pathfinders who discovered the power of conditioning – Ivan Pavlov and B.F. Skinner – did not give us training plans as such, but once we understand the principles they describe, we are able to devise our own plans. So, there is no '*Simon Prins' training method*', but there are '*Simon Prins' training protocols*', which detail the strategies I have used to teach specific tasks. I have learnt so much from what has gone before, as well as drawing on current research. My intention is, therefore, to encourage trainers and guardians to understand the basic elements of science and incorporate it into their work. Discover what has already been discovered! Engage with the scientists! In this way our training crosses new frontiers, becoming more efficient and effective and, most important of all, more compassionate in relation to our dogs.

CLASSICAL CONDITIONING (CC)

Pavlov and his team of scientists.

Ivan Pavlov's discovery of classical conditioning was accidental. As already described (see *page 24)*, when his team of scientists, dressed in white lab coats, were working with dogs to induce saliva production, he noticed that they salivated in anticipation of getting the food, and this was triggered by seeing the scientists in their white coats. Many of us are familiar with Pavlov's foundation work, but let's take a closer look so we can understand its full significance.

When we talk about classical conditioning, we need to examine reflexes. Reflexes are the internal and psychological responses of an organism to its environment. We all have reflexes – we need them to survive. However, we rarely think about them; they are just there, switching on automatically when we need them. For example:

- If a fly approaches your eye, you involuntarily close your eye to protect it.
- If you touch a hot surface, you involuntarily pull your hand away to protect yourself from serious burns.
- If your lungs are measuring too much CO2, you exhale and inhale.

Classical conditioning is reflex; you cannot stop it, nor is it easy to change it. If someone blows air in your eye, can you stop your eye from closing? Classical conditioning also has a biological basis that is linked to survival. It triggers reflex reactions such as salivating, blinking, avoiding pain, digestion and breathing.

When Pavlov studied the salivary reflex in dogs, he noticed something he didn't understand. Initially, he simply measured the amount of saliva a dog produced after being given meat powder. However, prior to the experiment, the dogs, who were waiting in another room, would start salivating, first on seeing the scientists in their white lab coats, and, in further tests, when a scientist approached with a harness. There was no food to trigger this reaction. So, unintentionally, Pavlov discovered what was later called a conditioned reflex (CR), now known as classical Pavlovian conditioning.

Pavlov showed us that it is possible to combine a neutral stimulus (NS) with an unconditioned – innate or natural – stimulus (UCS) to obtain an unconditioned – or innate – response (UR). Let's look at an example to better understand this scientific description:

A **neutral stimulus** (NS) is a stimulus that, by itself, does not elicit an observable change in behaviour. The sound of a clicker means nothing to an untrained dog. We see no reaction.

An **unconditioned stimulus** (UCS) is a stimulus such as a piece of chicken. Even if the dog has never tasted or smelled chicken before, he will have an **unconditioned reaction** (UR) if you hold it in front of his nose. We can observe that the dog begins to salivate.

Pavlov's work tells us that a neutral stimulus (a clicker), can elicit an unconditioned response (salivation) if a piece of chicken is given immediately after the click. This process is known as 'conditioning' or a 'conditioning process'. Once the conditioning process is complete, the scientific terms change. From now on we call the formerly neutral stimulus (NS) a conditioned stimulus (CS) because, even without the presence of chicken, it can elicit a conditioned response (CR), in this case, salivation.

If we break down what has been described into a simple table, we can see the individual steps very clearly:

Step 1: Sound of the clicker (NS).
Step 2: Food is presented (UCS).
Step 3: The clicker becomes the conditioned stimulus (CS) – the dog thinks food is coming.
Step 4: The clicker elicits a conditioned response (CR) – the dog salivates.

Once we understand these four steps, it is easy to grasp a simple classical conditioning process and apply it to dog training.

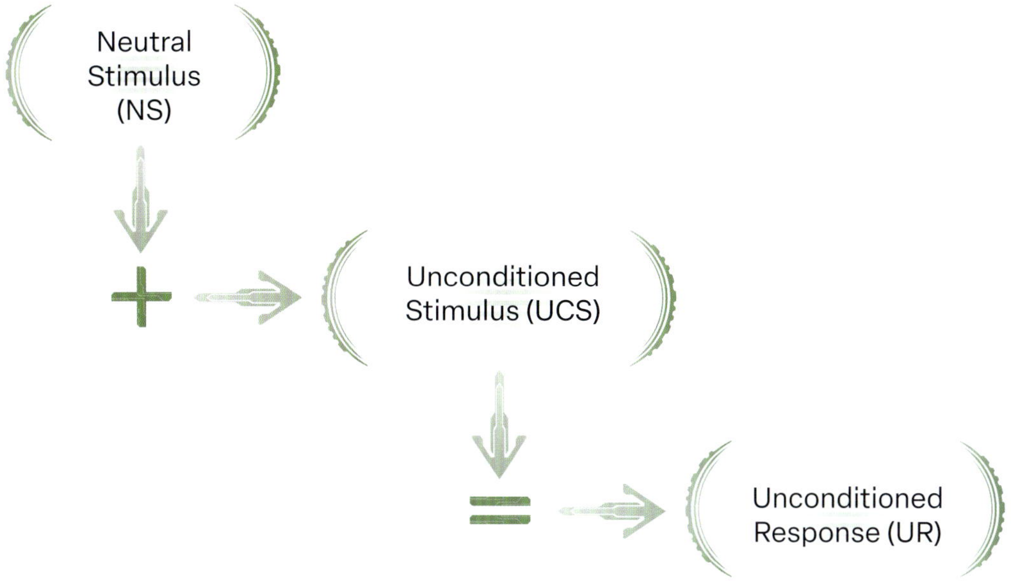

The beginning of the classical conditioning process

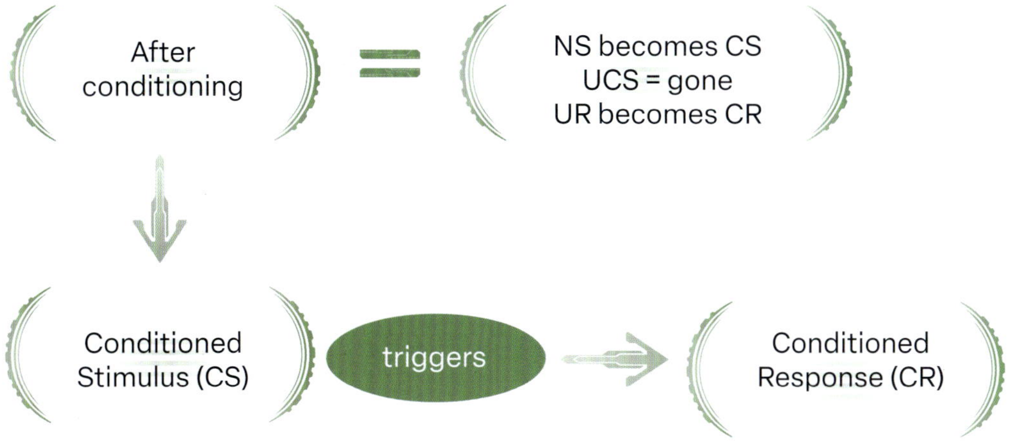

Process of classical conditioning completed

The sound of a clicker only changes from a neutral to a conditioned stimulus when it is followed by a primary reinforcer, such as food.

Pavlov found that he could transform almost any neutral stimulus, such as a sound, light, smell, or visual signals, into a conditioned stimulus (CS). By combining the neutral stimulus (NS) with an unconditioned stimulus (UCS) such as food, the neutral stimulus (NS) becomes important to the dog. The bell, clicker, white coat, harness, or a specific smell are no longer neutral but become a cue to the dog that something good is coming (UCS). After a few repetitions, the dog begins to anticipate the neutral stimulus, thereby completing the conditioning process. The neutral stimulus (NS) therefore becomes the conditioned stimulus (CS). The dog's response is no longer unconditioned (UR), but conditioned (CR). The key element here is that the unconditioned stimulus (UCS), such as food, is no longer present.

Classical conditioning is very powerful; we cannot 'turn it off' during the training process. It can be hugely advantageous, especially when it comes to conditioning target scents, but it can also slow down training, or result in the dog learning unwanted behaviour. For example, a common problem when training scent detection dogs is that the dog begins to indicate packaging materials, such as plastic bags or storage containers, or the specific human scent of the assistant who has set out the search items – or even the scent of the gloves worn by the assistant. All of these 'new' smells, which are unrelated to the target scent, can become new stimuli for the dog, and before we know it, the gloves have become a conditioned stimulus (CS). The dog no longer looks for the target scent; the smell of the gloves triggers a conditioned response (CR) because he is anticipating a reward.

"Pavlov is always sitting on your shoulder..."

Pavlov conducted a series of experiments to better understand the principle he had discovered by accident. In addition to using neutral stimuli, such as light, various noises and other triggers, he also experimented with time intervals. He tried giving the meat meal prior to introducing a neutral stimulus (NS), but that didn't work. As a result of running tests, collecting data, and experiencing success and failure, he reached the conclusion that all types of neutral stimuli can be conditioned, and can then be used to trigger a specific behaviour. This is a principle that is key to our training today – a ball, a kong, a laser, a bitearm, a scent, or a verbal cue are all perceived as 'neutral'; it is the process of classical conditioning that gives them value. Pavlov also discovered how subtle changes in behaviour can be introduced through the following processes:

EXTINCTION

This is the term given to a behaviour that fades, generally through lack of use, and is then extinguished. The reason we continue giving reinforcers after a bridge signal, such as a 'click' or a clicker word, is the danger of impending extinction. If the conditioned response (CR) is no longer combined with the unconditioned stimulus (UCS), the conditioned response (CR) gradually weakens and eventually disappears.

GENERALISATION

In his experiments, Pavlov discovered that the conditioned stimulus (CS) is quickly generalised, i.e., it becomes widely applicable. So, if the initial CS was a sound with a frequency of 1000 cps and then a sound of 1500 cps was presented, the conditioned response (CR) would also happen in response to the higher frequency.

We use generalisation all the time in training. Let's take conditioning a dog to a new scent as an example:

- Eight containers are placed in the form of a line-up. Seven of the containers are empty, and one contains the new target scent.
- The dog sniffs all the containers and, as soon as he sniffs the target container, he is reinforced with food.
- After three to five repetitions, the new target scent (NS) is associated with a UCS (food). Now the target scent is no longer a neutral stimulus (NS) but has become a conditioned stimulus (CS).
- When we have conditioned the new target scent, we can swap it with a new odour. Again, seven containers are empty and only one contains the new scent. The dog will quickly generalise and indicate the new scent, knowing that this behaviour will be rewarded.

The ability to generalise allows the dog to build up a library of target scents. However, it can also lead him into making mistakes. As we have seen, the scent of the gloves worn by the assistant setting out the containers is also a conditioned response (CR). This applies equally to other cues, such as verbal commands and body language. Before we know it, the dog has generalised words and signals - often given unintentionally - and they trigger a conditioned response. Generalisation can help us or work against us. Bear in mind, classical conditioning is very powerful, and it happens very quickly!

The dog picks out the container with the target scent and is rewarded.

DISCRIMINATION

Returning to the experiment with sound frequencies, Pavlov discovered that if he *only* reinforced the sound when it was at a frequency of 1000 cps, the conditioned response (CR) to the 1500 cps frequency gradually weakened and disappeared (*see page 56).* Therefore, if we get a conditioned response (CR) to the scent of gloves in our search area, if we don't reinforce it, it will go away. The dog learns that the smell of gloves alone does not lead to a reward. However, he may think that the target scent *plus* glove smell is the reward - and not target scent alone - so he may ignore it. Never underestimate the power of classical conditioning! Designing accurate protocols is the best way of avoiding such issues, as well as being hyper vigilant for signs of them occurring in training. The same applies to verbal cues: if we only reinforce *"sit"* - and not *"lit"* or *"wit"* - the dog will only respond to *"sit"*.

INVOLUNTARY LEARNING

Classical conditioning is often described as 'involuntary learning' because the conditioning process takes place even if the individual is not aware of it, or maybe does not want to learn anything. We cannot stop the process. Trainers of scent detection dogs also make use of involuntary learning by using a cut-off piece of a kong when training a new scent. If the dog knows the kong's odour and reliably indicates it, the kong's odour will be combined with a new scent After just a few repetitions, the dog will begin to indicate the new scent, even if the piece of kong has been discarded. The same principle is evident when a dog jumps up with excitement when he hears the doorbell. We have not trained this behaviour - the excited response is the result of involuntary learning, i.e., through classical conditioning combined with the stimulus of the doorbell - so for the dog, the doorbell means excitement.

Many detection dog trainers work with a piece of a kong to start training a new scent. The kong smell, which is familiar and has positive connotations for the dog, serves as a training scent, which is gradually replaced by the target scent. Classical conditioning at work!

OPERANT CONDITIONING (OC)

The difference between classical and operant conditioning can be difficult to understand, given that both are at work when we train. When training, we cannot choose between classical *or* operant conditioning; the two processes go hand in hand.

The best way to distinguish operant conditioning is to think of it as involving the striped muscles, which are used for conscious activities such as barking, running, jumping, fetching, sitting, lying down, and many other 'operative' activities. In other words, the dog shows behaviours that we can observe. Operant conditioning is also described as conscious learning. The individual may choose not to participate in the training process and, if this is the case, nothing is learnt.

Classical conditioning involves involuntary learning.
Operant conditioning involves voluntary learning.

Traditional trainers share a notion that operant conditioning relies solely on positive reinforcement, i.e., 'throwing treats', and, as such, it is not always taken seriously. This is unfortunate, because those who work with operant conditioning find it to be a very powerful technique, although it is a challenge to use it to its full potential. B. F. Skinner, a leading pioneer of operant conditioning, explained that if we want stronger behaviour and more of it, we need to use both positive and negative reinforcement. Conversely, when we want less of a behaviour or want to weaken it, we need to use positive or negative punishment. This resulted in the famous table of the four quadrants:

The four quadrants of operant conditioning

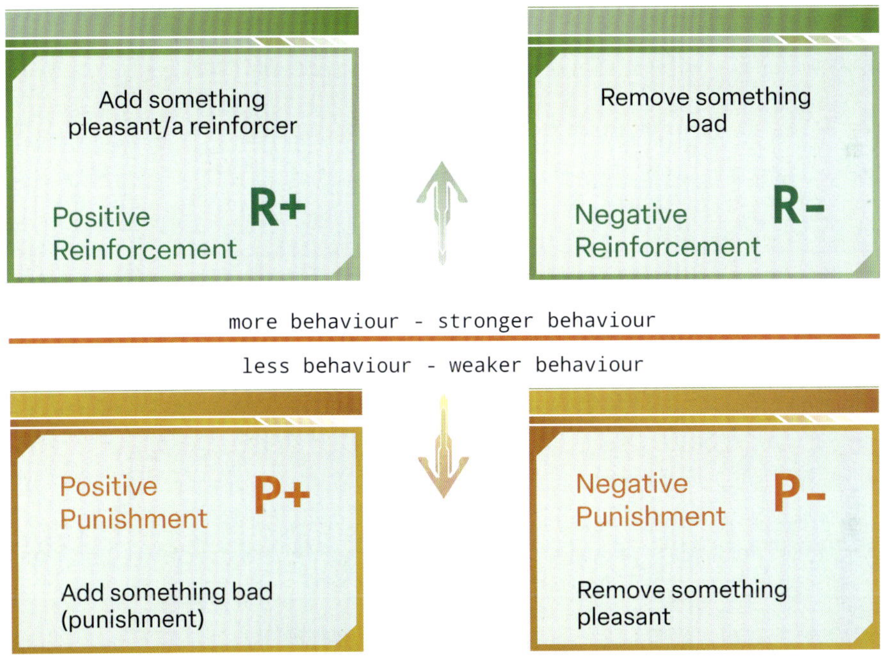

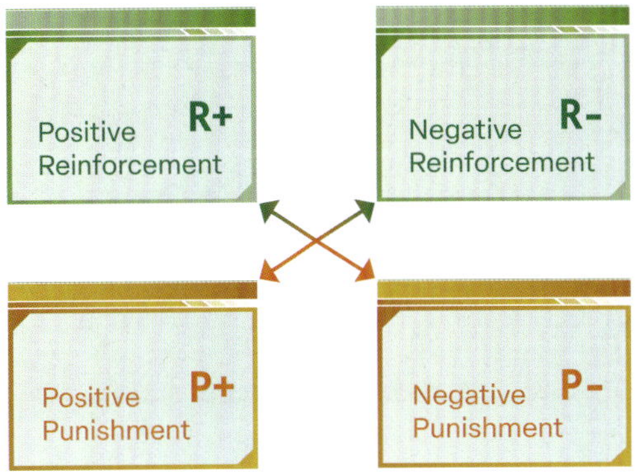

The four quadrants are not so sharply separated from each other as we may think. They interact and cannot be regarded in isolation. For example, in order to remove something bad, you must have put it there in the first place!

The dashed arrows between R+ and P- and R- and P+ make the cross-connection between the two clear. In order to be able to take away something unpleasant, you have to have added it beforehand or it must have been there before! Therefore none of the four quadrants are "in a vacuum")

TRAINING WITH THE FOUR QUADRANTS

Within the four quadrants lies the basic 'code' for training all living beings:
Positive reinforcement (R+) and negative reinforcement (R-) are used to get more of a behaviour, or to reinforce it.
Positive punishment (P+) or negative punishment (P-) are used to weaken behaviour.

Often trainers are not even aware that they are using all four quadrants, and many debate the difference between positive punishment and negative reinforcement. This following example gives us clarity:

- A simple command, such as *"No!"*, can be positive punishment if it means something bad is being *added.*
- A repeated *"No, no no!"* - creating mental pressure - can also become a negative reinforcement if it is *taken away* - relieving the pressure - when the dog performs the wanted behaviour.

The negative reinforcement starts *before* the (wanted) behaviour; the positive punishment *follows* the (unwanted) behaviour.

An important finding follows from this: whether something is a reinforcement or a punishment depends on the context, the point in time and, above all, the result. If the behaviour increases, it is a reinforcer; if it is weakened, it is a punishment. Contrary to popular belief, 'punishment' is not only defined by the dog being disciplined in some way, it also depends on what the dog perceives as being punishing. For example:

Animal training is simple, but not easy!

- A food reward is positive reinforcement; not giving a food reward if the animal does not meet the training criteria is negative punishment.
- A tug on the leash is positive punishment; releasing the line tension is negative reinforcement.

We may think we are only using positive reinforcement but, in all likelihood, the four quadrants are involved.

Whether something is a reinforcer or a punishment depends on the context, the timing and, above all, the result.

- **If the behaviour increases, it is a reinforcer.**
- **If it decreases, it is a punishment.**

This makes it crystal clear that operant conditioning does not only consist of positive reinforcement. Remember, behaviour changes because of its consequences. While we structure our training around positive reinforcement, we also use the other three quadrants when necessary for the safety of the individual animal, for other animals and people present, for the sake of the operation, or to prevent any other environmental danger.

Armed with the four quadrants and the basic principles that Pavlov gave us, we have everything we need to start training. Now it's simply a matter of thinking and planning, and putting it all into practice – think, plan, do! As Bob Bailey's reminds us: *"Animal training is simple, but it's not easy!"*

Classical or operant?

Attempts to distinguish whether a training method involves classical or operant conditioning always provokes lively and lengthy debate. Of course, I am in favour of discussion, but it is easy to get bogged down and to fuel polarisation within the training community. I cannot help but see this as a waste of valuable training time. I am more in favour of keeping it simple, so let's use an example that we can all understand:

As soon as a dog sees that you are cutting up treats ready for training, he will get excited, bark, jump up and salivate. To understand what's going on, consider this difference:

Classical conditioning involves reflexes, biological or physical states.

Operant conditioning involves voluntary activities, such as barking or jumping.

THE LEGACY

Pavlov and Skinner gave us valuable insights. Pavlov taught us that classical conditioning triggers reflex behaviour which is difficult to stop:

The dog is asleep in the living room when the doorbell rings. He does not respond and so his guardian opens the door, happily allowing the visitor into the house. The dog jumps up, barks and then gets lots of attention and praise from the visitor. Following a number of visits, along the same lines, the dog jumps up and barks every time the doorbell rings. After a while, the guardian is seeking help because the dog is barking frantically whenever the doorbell rings, and it is driving the neighbours crazy.

Skinner made it clear that learning is a process of both positive and negative:

- If we want to increase motivation – make the behaviour stronger, faster, more frequent – we use **reinforcement**. There are two forms:
  - Positive reinforcement (adding something nice).
  - Negative reinforcement (taking away something bad, e.g., pressure).
- If we want to get rid of a behaviour – weaken it and eliminate it – we use **punishment.** There are two forms:
  - Positive punishment (add something the animal doesn't like)
  - Negative punishment (take away something the animal likes).

With regard to using all four quadrants, the key lies in the time-lag between the dog's behaviour and the handler's/environmental response. Does the response trigger a change in behaviour? If the dog's behaviour does not change, we need to take a step back and create a training plan. Training must focus on behaviour change!

All trainers who use penalty-based methods also use positive reinforcement. And believe it or not, all 'positive' coaches use punishment, too. Often, we are simply unaware of what methodology we are using:

- The positive trainer's dog, who suddenly

pees in the living room, will most likely get a punitive "*No!*".

- The penalty-based trainer who pets their dog after a penalty is using positive reinforcement (as long as the petting is perceived as being positive).
- The positive trainer who fails to click/mark a behaviour, even though the dog has responded correctly to a signal, is punishing the dog by not giving him a reinforcer.
- The dog who would like to free-run with his friends, but is restrained by the leash, is enduring a 'punishment'.

From these examples it is easy to see that there is a thin line between reinforcement and punishment; something that is 'positive' can have a negative impact - and vice versa.

WHY IS OPERANT CONDITIONING SO POWERFUL?

By Marian and Bob Bailey

Behaviour analysis is an academic term for scientific study and changing behaviour using both operant and respondent conditioning. We will use a more popular term, operant conditioning, rather than the broader and less common, behaviour analysis. Operant conditioning is *not* a theory; it's a scientific, naturalistic account of how behaviour changes.

Operant conditioning works for us: The primary reason we believe in and use behaviour analysis is empirical; it works! It has worked for both of us for a combined period of more than 100 years of shaping the behaviour of more than 15,000 animals representing over 140 different species, from cockroaches to killer whales. We have controlled animals over great distances – seagulls and dolphins over many miles, dogs and cats over hundreds of yards. We have trained using sophisticated electronic equipment; e.g., lasers, radar, global positioning systems (GPS). But we have also trained wild raccoons simply by the timely throwing of pieces of boloney sandwich. We have never experienced animals being bored or wilful. We have never experienced a failure of the principles of operant conditioning – those principles discovered by B.F. Skinner in his laboratory.

It works for others: Where operant conditioning has been effectively used, it has worked. That includes not only training animals and non-verbal humans, but also changing complex human behaviour, verbal behaviour, education, business, safety, mental health, sports training, the arts... the list goes on and on. We have gathered on videotape the operant conditioning experiences of hundreds of scientists, physicists, chemists, zoologists, biologists, physiologists and even 'cognitive' psychologists. We have also interviewed teachers, bankers, business people, engineers, manufacturers and others. Not once did any of them report that operant conditioning principles failed them. They did find that they sometimes failed to correctly, or appropriately, apply the principles. In a complex behavioural programme, such errors are understandable.

Non-reasons for belief: We do not consider B.F. Skinner, nor any other person, infallible. Our reliance on operant conditioning does not depend on blind loyalty or on the commandments of others. Our belief is not based on our speculations, nor the speculations of others. Our belief is not based on anecdotal evidence, though we find significant casual observations (our own and others) very interesting, and sometimes useful.

Operant conditioning is based on science: We are biased. We believe that behaviour is determined. While behaviour and its causes may be very complex, they are not mystical or beyond understanding. Neither humans nor animals are machines. Humans and animals make decisions based on past experience and their internal and external stimuli. We believe in data, and the objective and cautious data analysis and interpretation of those data. Our adherence to the scientific method does not preclude us from speculating, nor from acting on our speculations. However, when we speculate and we make mistakes, we are prepared to pay the price of wrong guesses, and not resort to rationalisations to explain our failure. Also, our application of science and technology is not perfect. Human error is always a factor in any application or endeavour.

We are confident of the power of operant conditioning: Some might confuse our confidence in our technology with an attitude of smugness. We do not feel that we are superior to others. We believe we are using the best technology to shape behaviour. When we teach, we teach confidence in operant conditioning. We do not use any other system for changing behaviour.

We will use the system scientifically proven best to change behaviour: Operant conditioning has been scientifically tested and demonstrated to work. We will change to another system for altering behaviour when a better system is found. We feel no special loyalty towards operant conditioning. The new system for changing behaviour must be proven scientifically that it is faster, more precise, and at least as safe for the practitioner and the subject.

Why do some trainers have problems with operant conditioning? All technologies have limits. Any method for changing behaviour is subject to human error. An operant conditioning practitioner must know the limits and the powers of the technology. He or she must be proficient in operant conditioning knowledge and mechanical skills. Operant conditioning must be used exclusively, and not as 'just another tool'.

Why doesn't everyone use operant conditioning? We don't know. We can only speculate. Perhaps many practitioners are content with the results of their training methods. Maybe there is a feeling of loyalty, either towards their methods, or towards some person or an idea. Some say operant conditioning is too strong; that it is 'mind control'. Other say it is too 'soft', that there is not sufficient behaviour control. Still others bring up negative and obscure ethical issues. None of these judgements are correct, in our view. Perhaps an individual should not judge until he or she has mastered and used operant conditioning skills.

Chapter 7:

WHAT DO WE MEAN BY 'TRAINING'

Training is changing behaviour – pure and simple. There are many reasons why people want to train dogs and change their behaviour. So, firstly, we should decide why we want to train a particular dog and, crucially, what we want to achieve.

We need to ask the following questions:

- Do we simply want a well-behaved dog?
- Are we interested in dog sports – in which case is the training on a mental or a physical level?
- Do we need the dog to assist in our work?
- Do we want the dog to work more with his eyes and ears?
- Do we want to focus on his sense of smell?
- Do we want him to employ all his senses?
- Are we training to improve the dog's quality of life? A rescue dog, for example, is easier to adopt if he knows a few basic commands and is taught to live in harmony with other dogs.
- Is the dog destined for the police, the military or for border patrol?
- Is there a problem that needs to be remedied, such as teaching a dog to relax and settle when he is home alone?

To change behaviour, we must start by making a list of wanted and unwanted behaviours. This will help us to prioritise what we want to train. It's like constructing a house: we must start with strong foundations before we start building the house itself. We need to:

- Establish communication.
- Work on trust.
- Make a top five list of reinforcers (see *page 115*).
- Observe behaviour.
- Collect and analyse data.
- Draw up a training plan – and follow it!

Training is empowering not only for dogs, but also for trainers. Everyone who works with animals faces problems from time to time, and it's wonderful to see a trainer grow in confidence when all the pieces of the puzzle eventually start fitting together. The unwanted behaviour disappears, the training skills and opportunities increase, and an enriching human-animal partnership develops. On many occasions, I have often seen trainers make real progress by stepping out of their comfort zone and learning new skills. These are important life lessons that go well beyond the realm of animal training.

Think about what happens on a talent show on TV. In the first selection phase, the candidates are often shy and stand with sagging shoulders. But once they get through the first round, they start believing in themselves. They change! It is not only the hairstyles, clothes and make-up that are different, they hold themselves differently – chest out, shoulders back. Their body language changes and they face the world with confidence – so much so, they seem like a different set of people. We can observe the same process with trainers. Many start out as handlers, kennel staff or dog walkers. Shy and inexperienced, they soak up any training information that they come across. They start walking their dogs differently and changing aspects of their behaviour. Their work improves, and they achieve results more quickly. They receive compliments from more experienced trainers or from management, and their confidence grows. They make fewer mistakes and begin to understand dog behaviour on a deeper level.

I observed this happening when I was working with new recruits in military and police special units. Initially, everyone was quiet, not knowing what to expect. They made mistakes, they had doubts, they felt vulnerable. But after their first boot camp, they started to change: they became blood brothers, they learnt to fight and operate machines. They developed new skills, their self-confidence grew. And let's not forget the dogs we trained. Every time we started training a new dog in the special unit, we asked ourselves: "*Can he do it?*". Because, initially, he didn't behave like the experienced, fully-trained dog that we had been working with. The new dog was slow in responding, and did not even understand *"sit," "down," and "here"*. We had to start from scratch and train these basic cues, along with many other skills. In the beginning we used a high/continuous rate of reinforcement (see *page 109*) for the wanted behaviour and asked for short duration. Our criteria was still broad. We worked close to the dog, ensuring body language was clear and offering help in the form of targets and platforms. Then we slowly began to taper off our assistance. We added more criteria, asked for greater duration and introduced variable reinforcement (see *page 109*). At this point we could see if our

BEHAVIOUR CHECK

AGGRESSION VERSUS ASSERTIVENESS

To train effectively, we need to develop an understanding of behavioural traits, particularly those which appear to be similar. It may appear that a dog who is assertive, and a dog who is aggressive, are one and the same. However, the difference between these traits is significant.

We see a lot of aggression in police and military dogs in particular, but aggressive dogs are not always the best dogs to work with. They can be unreliable and unpredictable in response to bystanders, and even to their handlers. There is always a reason for aggression. It can be caused by problems in training, it can come from hidden pain or frustration, or it can just be a lack of clear boundaries. Working with aggressive dogs can lead to dangerous situations, especially if it involves teams with other dogs.

However, we do want assertive, determined dogs for our work. Dogs who are confident and mentally strong are very reliable. They can work in stressful situations, remain calm and do their job. In operations, I have often seen aggressive dogs make critical mistakes, such as biting their teammates, or attacking the wrong person. In contrast, confident and assertive dogs seem to understand who is bad and who is good; they have the ability to problem-solve.

BEING TOGETHER

The more time you spend with your dog, the more you will learn from each other. There are so many subtle signals and clues that tell you more about each other. Dogs are social creatures and enjoy closeness. However, there are dogs who are 'excluded', spending lengthy periods of time on their own: at home, in a car, or in a crate or a kennel.

How far does 'inclusion' go with you? Trust comes from spending time together and sharing activities, which should not be confined to training. Plan little adventures, explore together, play – just be together. But remember, a dog also needs to rest, especially after training when he needs to process what he has just learnt.

This is inclusion: A training break in the living room.

training was effective - had we succeeded in changing behaviour? Had the dog gained in confidence?

The best evidence came from watching young search dogs working over an area of rubble for the first time. They moved slowly and carefully, keenly aware of the strange, unstable ground. But after just a few repetitions, they were moving at speed, keeping their balance, avoiding injury and crawling into the smallest holes in search of survivors. We could see the transformation with police patrol dogs. When a young police dog attacked a 'suspect' for the first time, his handler helped him by keeping him on a short leash and standing close to him. The dog was praised for biting and encouraged to keep attacking. A few months later, the handler was able to unleash the dog a few hundred metres from the suspect. The dog then took off in hot pursuit, launching into the attack with a mighty leap. He was full of energy and showed no fear.

Training is not just about changing behaviour. It is empowerment of the individual, giving them the skills to do their job, and encouraging physical prowess, courage and confidence. Training can be a life-changing experience!

DISCIPLINE AND CORRECTION

The need for obedience – and for discipline – is a sensitive subject not only in the world of animal training, but also in our society. The 2020 pandemic has shown us how difficult it is to get people to follow rules. Why is it that some people respect rules, and others flout them? In Asia citizens wore masks from the moment it was ordered by the government, but in Europe this rule was challenged and it was a long time before people complied. Meanwhile, citizens in the USA voiced the importance of freedom and fundamental human rights, and many refused to wear masks or go into lockdown.

Obedience, submission, rules, and discipline can help us survive, stay intact, and work more effectively and efficiently. For example, a few weeks ago I witnessed a dog being hit by a car. The dog was out for a walk with his guardian when he saw another dog across the street. He ignored his guardian's attempts to call him back and ran into the car. He died from his injuries and his guardian was devastated. On another occasion, I witnessed a large dog attack and kill a much smaller dog. The guardian of the dog who was attacked tried to intervene and was also bitten.

Regardless of whether we are on the road, in the air, working with potentially dangerous machinery, or caring for zoo animals, we accept that compliance and discipline are necessary. In the military, especially during selection and basic training, we know that discipline separates the weak from the strong, the aggressive and insecure from the confident. Discipline is not a bad thing.

However, the use of the word 'correction' triggers strong emotions as, in our minds, it is linked to physical correction. We are conditioned to think that correction is something negative that damages relationships. However, this does not need to be the case. In the learning process, corrections are a necessity; learning without corrections is simply not possible. The problem is that many people do not understand the definition of a correction. So let's get some clarity on this with a dictionary definition:

A correction is the act of offering an improvement to replace a mistake.

With this definition we understand that withholding a reinforcer, if criteria is not met, is a 'correction'. Let's take the exercise of teaching a dog to sit, as an example. If we cue our dog to *"sit"* and our training plan expects the 'sit behaviour' to occur within two seconds, we will not reinforce the act of sitting if it takes the dog longer than two seconds to respond. We will not give him the treat or toy because we did not get the behaviour we wanted. Not giving a reinforcer is a form of correction which is used to get the dog to give us the behaviour we want. But, of course, we all make mistakes – not just the animals, but the trainers, too. If corrections are too harsh, it is often because training did not go as expected. In many situations, overly harsh corrections are the result of frustration.

This can be resolved if we work with a training plan, noting down all the steps we need to take. We build our plan in small, incremental steps so that success is virtually guaranteed. As long as we can control the environment, we don't need any 'fixes' other than a restart. A restart means that the training attempt is terminated and the animal can no longer earn a reinforcer. He must repeat the attempt from the start.

As training progresses, other corrections may be required. We must prepare the dog for the challenges that life will throw at him, working with a variety of distractions he may encounter. He may come across something he wants to chase, or he may find some discarded food – these distractions crop up in operations, in competitions, and in everyday life. Corrections can be verbal, signalled with body language, or with a long leash. This has nothing to do with mental or physical abuse. The goal is to change the behaviour and add a negative consequence to ensure we can keep the animal from harm. Corrections need to be precise, just strong enough to change the unwanted behaviour, carried out at exactly the right time, and executed without emotion. Once emotions are involved, correction becomes a completely different issue and there is a danger that feelings will run out of control.

THE CONCEPT OF DOMINANCE AND HIERARCHY

Is it legitimate to compare dogs to wolves? Are the rules of hierarchy the same for both species? There are different views on this, even among scientists. Whatever we choose to believe, we must remember that we may also encounter aggression related to dominance and hierarchy issues. The fact is that both exist and can cause problems. Dogs can be aggressive towards other dogs, but also towards people or even their guardians.

The concept of dominance has been hotly debated in recent years and is the subject of many books. In summary, yes, there is dominance among dogs, but it is not as solid and linear as once thought. It is not a character trait in the sense that a dog is *always*, or *attempts* to be, dominant over other dogs, or humans. It is a matter of dominance relationships changing according to circumstances, context, or the presence of other animals or humans.

BEHAVIOUR CHECK

NATURE OR NURTURE

This is the age-old question: which part of a behavioural repertoire is already laid out in the DNA or the genes (nature), and which is learnt from experience (nurture).

Dogs learn by experience. A dog who plays too close to the bank and falls into the water will learn from the experience, i.e., he will adapt his behaviour if he meets with similar situations in the future. I once witnessed an interesting scenario when two of my dogs were chasing each other around my parked car. If they were on opposite sides of the car, they would both stop. The older dog would then look under the car at the youngster's feet. If he saw him moving to the left, for example, he would head off in the opposite direction to prolong the chase!

However, there are some behaviours that are firmly embedded in the DNA. Without experience or training – Border Collies will naturally herd sheep and Pointers will 'freeze' when they smell a bird in the field.

There is also a hierarchy among a team of sled dogs.

As already highlighted, dogs are social animals, and they enjoy living with others.

However, there can be serious issues relating to dominance aggression and, according to Peter L. Borchelt, who made a study of aggressions in dogs, it can take the form of highly complex aggressive behaviour along with other behavioural problems. In an article entitled *Dominant Alpha Humans Do Not Gain Dogs' Respect and Trust,* behaviourist Marc Bekoff puts forward the thesis that dogs form dominant relationships within functional packs, and there are alpha individuals.

In countries where there are still large populations of free-roaming dogs, it is evident that the dogs form functional packs as communities of convenience, and this helps them to survive. Often, there is no fixed alpha pair at the top of the hierarchy, as is the case with a wolf pack. Instead, the dogs take turns in the distribution of tasks (and leadership) depending on which individual is currently contributing the most to the wellbeing of the group. This model of the 'functional pack' may come closer to an ideal human-dog relationship in that humans should be willing to hand over responsibility to their dogs if they can do something better - for example, following a scent trail. However, many people find it difficult to relinquish control!

The term 'alpha' confuses many people because of its association with the negative effects of alpha-oriented and fear-punishment-based training. However, it is not synonymous with being mean or aggressive, or any other traits that are perceived as negative. The goal of the alpha individual in a group is to create a solid structure for survival, to set some rules and thereby provide order and stability. Aggression and fighting could injure the alpha, which is not conducive to pack survival. The alpha is more likely to ensure that there is cohesion in the group using non-aggressive and non-violent methods. He asserts his position not by fighting, but by asserting himself.

This does not mean that a dog living in a human household sees his people as a 'pack'. There is no data or research to support this thesis. As such, there is no evidence or justification for the claim that humans should be alpha to dogs. It is not necessary to dominate or force a dog into submission to change his behaviour. Dogs cannot understand why they are loved by their people one moment and punished the next. Positive training is the better solution. It may be more challenging than using dominance-based training methods, and it may take longer, but it will definitely improve our relationship with our dogs. *For more information on dominance and hierarchy, see Chapter 22, page 244.*

Chapter 8:
METHODS OF TRAINING

There are many different so-called 'training methods' and the number is increasing as trainers claim they have come up with a whole new approach, although very often this is nothing more than a combination of existing methods. I have selected the best-known methods that are frequently discussed in workshops and seminars, and those that need a little more explanation.

As previously highlighted, there is no *Simon Prins training method*; in fact, there is no such thing as a 'training method' because *all* learning is based on classical and operant conditioning. Behaviour changes because of its consequences! When behaviour changes, there is always a reason for it, i.e., through the introduction of either a reinforcer or an aversive. This process may not be easy to recognise at first glance (intentionally, or otherwise) which leads trainers to claim that that are *"communicating not conditioning"*, or they are working on *"relationship building in place of conditioning"*. Don't be fooled! On closer inspection, we will *always* find the laws of operant and classical conditioning in action, simply because that's the way learning works. Often, positive reinforcement training is equated (or misunderstood) with *"feeding the dog all the time"* which is entirely misguided. I hope this will become clear as the book progresses.

Instead of discussing different 'training methods', we should focus on the four quadrants (see *page 59*) and work out how they are being employed, and which are prioritised. Does the 'method' involve rewarding a lot or punishing a lot? Is the omission of pressure defined as a reward? When evaluating 'training methods', I recommend the following exercise:

Identify the reinforcers and aversives that change the behaviour!

In that way, you will truly discover which of the four quadrants are at work.

There is no such thing as a 'training method' that does not involve operant and classical conditioning.

TRADITIONAL PUNISHMENT-BASED TRAINING

It is difficult to give this type of training a specific label, so I have grouped similar training methods into one category. What all these methods have in common is that they have not been validated by scientific research, which is one of the reasons they are fiercely contested. However, they exist, they are used, and, in some ways, they are successful. Generally, this type of training is based on coercion. Trainers 'punish' any type of behaviour they don't want to see, leaving only the desired behaviour in the dog's repertoire. There are no books or written protocols; the method is passed on orally and students learn from 'experienced' trainers.

To explain the principle of how this type of training is structured, I often use the example of a taxi ride. If I flag down

A PARADIGM SHIFT

When I entered the world of police dog training in 1993, there was still a great deal of force-based training. Punishment seemed to be the basis of all training, which was a surprise to me as I expected these 'professionals' to use different methods. I was forced to adopt coercive methods because, at that time, it was the only way to train general purpose police dogs. At least that was the opinion of the chief instructor and the rest of the group. After a while I came to understand that this narrow-minded approach was caused not only by lack of knowledge within the group, but also because there were no training standards set by any organisation. There was no formally regulated training path, and the training process was based on assumptions and beliefs (paradigms).

As I described at the beginning of this book, when I started training police dogs, I was already against using coercive methods. In fact, I was in danger of being fired as, according to the chief instructor, I wasn't "tough enough" for this type of work. However, I stuck to my guns and started reading and researching, and just training differently. In time it paid off: my police dog was working well out on the streets and our performance in the field was excellent, and this was a direct result of reinforcement-based training. The relationship between me and my dog improved dramatically as we learnt to trust and respect each other. My dog felt empowered and this had a very positive effect on his work. We were sent on the most challenging operations where others didn't dare send their dogs. This eventually brought about a paradigm shift in our training group and other handlers began to use reward-based methods as well.

a cab, get in the back seat, and don't say a word to the driver, he will probably just pull away. If he opts to turn left at the first intersection and I smack him in the face, he might turn right. That way, maybe someday, we will get to my chosen destination. By hitting, or not hitting the cab driver, I get to my goal.

However, this is not not necessarily the way we would like to communicate with humans, or animals. It's not effective, it's not efficient, and it doesn't build good relationships.

AVOIDANCE-BASED TRAINING

This is a quick, easy and effective way to train, and like all other training methods, it works not only for animals, but for humans as well. The premise is that adding and removing pressure can influence behaviour. A puppy may freeze and plant all four paws on the ground the first time he is asked to walk on the leash. He may then try pulling on the leash, but this results in pressure on his collar. So, after a few repetitions, he will learn to walk without putting tension on the leash, as this relieves the pressure.

Making use of this response as a form of training, is highly dangerous, not only in terms of damaging the relationship between trainer/guardian and dog, but it can also be very harmful on a mental level for the dog concerned. If the dog loses awareness of what is causing the pressure, discomfort or pain and doesn't know what to do to avoid it, he will mentally and physically 'surrender'. Over time, he will no longer try to escape or avoid the situation, he will just suffer. It can only be deemed as cruelty if an animal (or a human) reaches the point where they cannot voluntarily leave a situation and must therefore endure it. *See Learned helplessness, page 135.*

CLICKER TRAINING

The clicker is a training tool – a device – that is used as a bridging signal. But there is no such thing as a 'clicker training method', despite what many believe.

The clicker is a neutral stimulus and, as such, it is meaningless to the dog. It is only after being paired with a primary reinforcer that has intrinsic value (such as food) via conditioning that it gains its power and becomes a signal that a reward is on its way. In the animal training world, the clicker is used as a bridge/marker and fits in the list of secondary reinforcers (see *page 78*). The clicker is not a magic tool; it is a tool that facilitates clear communication. When we use a bridge signal, it means that the animal receives a positive reinforcement (e.g., food or toy) after the click. So, the clicker bridges the time-lag between the target behaviour and the primary reinforcer. It is important to keep the communication clear and the expectations of the animal high.

There are different ways of using the clicker in training. Some trainers 'charge' it so it can be used multiple times in succession without having to give the animal a primary reinforcer, such as a food, or play. For example, after the click a piece of food is given. After repeating this 50 times the clicker itself becomes powerful. Every time the dog hears the click, he will experience a release of dopamine – the happy hormone – as he anticipates a reward. When this positive response to the clicker is strong, the clicker is charged or 'loaded'. So, when the dog is doing something good the sound of the clicker is used to highlight and reinforce that moment, but no primary reinforcer is given.

Of course, over time, the dog's response will weaken because the 'click' has not delivered a reward. A more effective way of utilising a charged/loaded clicker is to use it as a 'keep going' signal. For example, if we are training a duration sit in obedience, where there are a lot of distractions, we could click at intervals to tell the dog to 'keep going' and a reward will follow. *For more information on keep going signals, see page 156.*

There are paradigms around clicker training that can be misleading. This is what happened after a clicker training course in the Netherlands:

Following negative publicity concerning the police using coercive measures in dog training, those responsible wanted to change the culture so they sent trainers on a two-day clicker training course. This was to be a new dawn in police dog training – the clicker was going to be the solution to all training issues. Unsurprisingly, this do not happen. As a result, many police dog trainers virtually banned the clicker as, allegedly, it did not work on police dogs. Unfortunately, an opportunity was missed to educate the trainers in the correct use of a bridge signal and to encourage them to study classical and operant conditioning.

As I have emphasised, the clicker is not a magical training device. It's a cue, a marker, a bridge that will communicate clearly to the animal that a positive reinforcer is on its way. It is a brilliant training tool but, unfortunately, the hype surrounding clicker training leads people to believe that that the clicker itself is the answer to all training problems. When a dog is trained using coercive methods, if the trainer does not understand the power of positive reinforcement, or is unable to reflect on their actions, mistakes, training protocols, results and consequences, a clicker will not help. It is not like fairy dust. Changing behaviour is only possible if we understand that we need to change and, we are willing and able to change. *For more information on bridge signals, see page 89.*

The clicker is not a magic device, it's just a secondary reinforcer that is only effective if it is used wisely.

'DO AS I DO' TRAINING

This is the new name for imitation learning, which, broadly speaking, involves the dog learning by observing others. The trainer uses body language, allowing the dog to observe closely and then motivates him to imitate. This method is becoming increasingly popular and although it is often described as "new and exciting", it was, in fact, pioneered by psychologist Albert Bandura in the 1970s. His book, *Social Learning Theory*, explains how people learn through observation and modelling. In a recent study, Claudia Fugazza and Adam Miklósi underwrite this theory, stating that this technique is more effective than shaping behaviour.

Social learning is certainly very interesting, but to claim that it is more effective than operant conditioning is questionable. A scent detection dog may follow the trainer's gaze, and sniff at that exact spot, but have you ever seen a trainer sniff out an entire set of scent samples as an example for the dog to imitate? *For more information on social learning, see page 122.*

'POSITIVE'/'NO HARM' TRAINING

This is a rapidly growing group of trainers who oppose the use of any form of negativity in training. Don't get me wrong, I am not in favour of using punishment in training. But within a learning process there will always be some unpleasant influences that help to determine behaviour.

As an example, let's take a dog receiving a cue to "sit". He sits down but doesn't meet the criteria, i.e., he doesn't sit down quickly enough, or he immediately gets up again. The dog expects a reward for his sloppy sit, but the trainer withholds it because the dog has failed to meet the criteria. To facilitate learning, the trainer restricts rewards to when the dog responds quickly to the cue, or when the 'sit' is of longer duration. Lack of reward is an unpleasant consequence that can change behaviour.

There are times when even the best-trained dogs exhibit unwanted behaviour. The only way to change it is to remove the reinforcer and, if that's not possible, to introduce a negative consequence, such as a reset, which can be a highly effective training tool. A reset is used if a dog fails to meet the criteria, and the trainer responds by stopping the exercise, waiting a few seconds, and then re-starting. From a training perspective we close 'the window of opportunity' for the animal to earn a reinforcer. After a short 'time out', lasting just a few seconds, we restart and there is a new opportunity to earn reinforcement.

From this we can see that 100 per cent positive training – avoiding all negative elements – is a real challenge!

For more information on the use of 'punishment', see Chapter 17.

CHICKEN TRAINING

The first chicken workshops were run by Bob and Marian Bailey to show the power of operant conditioning. In recent times, chicken clicker courses have been advertised where the primary focus is on using the clicker. But here, too, the clicker is only the bridging signal to announce that food is coming and, thus, it is only a small component in the process. Participants on these courses are taught to lure the chickens around slalom poles or to complete an obstacle course, but they do not apply the principles of classical and operant conditioning to change behaviour.

Bob and Marian chose chickens as behaviour models to show that their training principles, involving operant conditioning, were valid for all higher animals. Their workshops proved invaluable for trainers, educating them and preparing them for the challenges they would meet in real-life scenarios. They learnt theory, and then put it into practice using chickens as live models. Training with live animals is very different from studying theory and working with chickens proved to be a real challenge as they move so fast and react to the most subtle of movements.

I completed all five stages of the Baileys' chicken workshops (and many more) which had a profound effect on my understanding of training and on my practical skills. I also learnt the importance of collecting and analysing data, as well as how to write a training plan. In addition, I learnt how to train coaches so, all in all, it was something of a life-changing experience.

In the chicken workshops conducted by Bob Bailey, trainers learn how operant conditioning works both in theory and practice.

BALANCED TRAINING

In addition to the 'positive trainers' there is also a growing community of trainers who are looking for a more balanced way of training, where there is no desire to work with punitive or coercive methods but, at the same time, accepting that all the answers cannot be found in 100 per cent positive training. These trainers promote boundaries, desirable and undesirable behaviour, corrections, and consequences.

VIRTUAL TRAINING PROGRAMMES

For those who are unable to work with real animals but want to improve their training skills, I recommend a piece of computer software – *Sniffy The Virtual Rat*. This programme was originally designed to enable psychology students to practise operant conditioning. Of course, it isn't a substitute for a real animal, but it comes very close to providing what we need to know to better understand animal training. It gives a sense of when repetitions are needed, what disappointment feels like and how animals interact with the trainer.

WHAT IS THE BEST TRAINING METHOD?

For me it is very simple: classical and operant conditioning offers the most varied range of possibilities. Trainers who are 100 per cent positive, limit themselves to using only one quadrant – positive reinforcement (R+). Balanced trainers use all four quadrants of operant conditioning.

Often trainers who claim to work exclusively with positive reinforcement (which, on closer inspection, does not exist, as we will see later) say that positive reinforcement training cannot be mixed with other forms of training. Once we start positive reinforcement training, we have to stick with it otherwise the animal will become confused. There is a grain of truth in that animals who have only experienced punishment-based training often fall into a degree of learned helplessness (see *page 135*). They would rather do nothing than try something that could lead to more punishment, which is a very sad state of affairs. In contrast, positive reinforcement encourages a willingness to experiment as the animal tries to 'earn' the next reinforcer. However, the belief that we can always work 100 per cent positive reinforcement is entirely misguided because if we are training – seeking to change behaviour – it is impossible *not* to punish.

As discussed, not giving a reward for failing to respond correctly to a cue is a 'punishment'. The dog is not fundamentally demotivated by this, as long as the pleasant experiences outweigh the negative experiences. Therefore, if a dog consistently fails to complete a training step, which means withholding the reinforcement/reward, we need to change our approach as quickly as possible so that he can be successful again. As far as the dog is concerned, more important than getting a reward ten times out of ten, is the predictability and 'fairness' of the trainer, which will impact on his understanding of the task. Failed attempts are part of the learning process and, without them, learning would not be possible.

For many people, the word 'punishment' is highly emotive term and fuels heated debate. Maybe it's time to change the terminology and use another word, such as 'correction' and steer the argument towards a more adult conversation.

Selecting the best method depends on differing scenarios, contexts, and the environments. We need to put our prejudices to one side and, instead of labelling all punishment as 'bad' we should consider the options. In some cases, the dog may suffer a vastly restricted life, or even be euthanised, if we did not seek to change his behaviour by making use of all four quadrants. Training animals of all kinds is a complex subject, and there is much we still have to learn. Let's hope research is fruitful and ongoing, allowing both trainers - and animals - to benefit in the future.

The best training method is the one that gives the animal the greatest possible chance of success.

Chapter 9:
THE KEY ELEMENTS

As a diver, I learned to rely on the basics – it ensured my survival if anything went wrong. Equally, the professional lives of pilots, doctors, fire-fighters, police and military personnel are based on key tenets of learning. These are the elements that hold everything together, in much the same way as bridges and buildings are constructed in such a way that they will not collapse.

When it comes to animal training, there are five fundamentals which are crucial to success:

- Stimulation
- Reinforcement
- Extinction
- Aversives
- Generalisation

Whatever we encounter – now or in the future – these five key elements should be our guiding light. We should always go back to basics! Extracted from the work of Thorndike, Pavlov, Skinner and the Brelands, they are the building blocks for conditioning new behaviour and changing unwanted behaviour. They have underwritten my thinking and my understanding, allowing me to communicate in a 'new' language, exchanging ideas with scientists, as well as with other trainers. Making a study of classical and operant conditioning and acquiring a good understanding of the five key elements – and how to apply them – will make us all better trainers!

Let us look at each of the five elements in turn.

Studies have shown that dogs can hear much higher sound frequencies than we do, so this type of sound, which is irrelevant to us, may be perceived as a stimulus.

STIMULATION

A stimulus is a change in physical energy that a living being can respond to, or sense. This includes sight, hearing, taste, smell, touch, but also movement (muscle and tendon sensations), temperature (hot or cold), pain and the vestibular senses, such as head movement and balance. The perception of stimuli varies from one species to another. We know that dogs and many other mammals rely heavily on their sense of smell, but they also have very good eyesight and hearing. Dogs can identify thousands of substances by smell. Touch and taste are very primal senses. Both animals and humans respond to tactile stimuli; the skin is our largest organ, and we not only react to touch, but also to temperature stimuli (hot or cold), pain, stretching and pressure.

If an animal cannot respond to a

given stimulus, then it is not a stimulus. For example, high-frequency sounds above 20 cps, are not a stimulus for us humans as they are outside our hearing range, but they are a stimulus for dogs.

REINFORCEMENT

Reinforcement is the process that strengthens a response. A reinforcer can be anything - food, an object, (e.g., a toy) or an activity - that strengthens a response. Reinforcement means a higher frequency of reaction. We, therefore, define the strength of a reaction in terms of the rate at which it occurs, or the increase in the likelihood that it will be shown. It is *not* about the intensity or severity of a reaction, but rather the frequency with which it occurs.

To select the best reinforcer, we must consider the individual animal. For some dogs, a tennis ball is the best reinforcer, while others ignore the ball but will do anything for food. It's like the smell of cigarette smoke: some people like it, while others hate it and walk away.

PRIMARY AND SECONDARY REINFORCERS

To provide effect reinforcement, we need to understand the difference between primary and secondary reinforcers. To keep it simple:

Primary reinforcers: These are things or activities that a living being needs for its survival, such as food, water, sex, shelter, warmth, social activities, etc.

Secondary reinforcers: These are things or activities that become valuable when they are paired with primary reinforcers. They include toys, balls, sounds, touches, etc. We also refer to secondary reinforcers as conditioned (learned) reinforcers or discriminative stimuli.

Reinforcers can be positive or negative:

Positive reinforcers that are given after a cued behaviour, strengthen the response. If a dog likes food and gets a treat after responding to his cue to sit, he is likely to show an increasingly strong response to that cue in the future. If this is the case, we can say that food is a positive reinforcer for the dog.

Negative reinforcers are those that avoid or terminate an aversive stimulus. For example, when a puppy is first put on the leash, he is reluctant to walk and may become even more resistant if he feels pressure on the leash. However, over, time, he stops resisting and the tension on the leash disappears.

While many modern animal trainers focus on positive reinforcement, we cannot deny that negative reinforcement plays a part in all forms of training. Negative reinforcement is effective. If we fail to fasten our seatbelts in the car, we have to listen to an unpleasant warning signal until we fasten them. Negative reinforcement causes us to move away from a fire if we find the heat uncomfortable. Likewise, a warning signal - a negative reinforcer - can stop a dog from barking if experience has taught him that something aversive will follow.

EXTINCTION

We can make use of extinction to get rid of an (undesirable) behaviour. If a previously reinforced behaviour is weakened, it will gradually fade until it disappears completely. For example, if a dog pulls on the leash, this behaviour will weaken and vanish if the handler stops each time the dog pulls and only continues when the leash is slack. Prior to this, the dog has been motivated to pull because it got him to the places he wanted to go. Now he understands that when he pulls, he has to stop, which means he cannot get to where he wants to go. Therefore, over time, his 'pulling on the leash behaviour' will weaken and eventually disappear. However, this will only happen if the handler is entirely consistent with the training technique of stopping when the dog pulls. If the occasional pull is tolerated – perhaps because the handler is distracted – the dog's pulling behaviour is reinforced and, is therefore worth repeating. This will have the long term effect of strengthening the undesirable behaviour overall. In this situation, the behaviour is being kept alive by an unintended variable reinforcement. *For information on reinforcement schedules, see page 108.*

EXTINCTION BURST

Trainers need to be prepared for this very interesting phenomenon when they are using extinction as a training technique. The unwanted behaviour that is targeted for extinction, may get even worse as training commences!

Take the example of the dog who pulls on the leash. When extinction training is underway – and the dog's behaviour is no longer being reinforced – he may pull even harder. It's akin to what we might do when faced with a car that won't start after turning on the ignition. Our likely response is to keep turning the ignition on and off – or even banging on the wheel – before we give up.

SPONTANEOUS RECOVERY

In our leash pulling example, it is possible that, after a few days of walking on a loose leash, the dog might suddenly start pulling again. However, if we apply the principle of erasure (extinction) the unwanted behaviour will disappear more quickly than the first time it was applied.

AVERSIVES

The use of aversives will weaken a behaviour; their application is referred to as 'punishment' – that emotive term that causes controversy among trainers and handlers (*see Shame and guilt in dog trainers, page 177*). However, as I continue to emphasise, aversives are part of the learning process. They help us survive. Bad weather forces us to seek shelter if we don't want to become ill and, potentially, fall prey to predators. If we are consistently late for work, we will be fired, stop earning money, and be unable to buy food.

We see 'punishment' in terms of an individual administering physical pain, or worse, in response to an unwanted behaviour. However, there are many other aversives that are the means of changing behaviour in both humans and animals. On a simple level:

- If we are burnt by fire, we learn to keep our distance.
- If it starts to rain, we look for cover.

However, our response may be more nuanced, influenced by context and experience:

- If we are in a hot country and there has been no rain for weeks, we may enjoy standing outside and getting soaking wet.
- A predator may get hurt as he hunts down his prey, but this is not sufficient punishment to stop the attack.

What is aversive, or punishing, also

depends on the individual.

- A well-trained police dog who is commanded to attack a criminal, may get injured. But he does not regard this as punishment as he has been trained to cope with it.
- But the Golden Retriever who steals a piece of meat from the kitchen table is very shocked when his guardian responds by hitting him with a rolled-up newspaper. He stops immediately and never steals from the kitchen again.

So, in general, an aversive is a thing or activity that provokes fear in a living being, and their response is to try to avoid it.

PRIMARY AND SECONDARY AVERSIVES

Like reinforcers, aversives can be categorised as primary or secondary:

Primary aversives are those that can cause physical harm or pain. Again, what is perceived as aversive depends on the individual. For some animals, it could be very loud or high-pitched sounds or certain smells that are unpleasant to them.

Secondary aversives are formerly neutral stimuli that have been combined with a primary aversive stimulus. For example, the word 'no' (formerly neutral) is combined with an aversive tug on the collar. Or raising a stick (formerly neutral), and then hitting the dog with it. Words, threats, body posture and warnings are all secondary aversives. Secondary aversives are a part of everyday life and are not always directly related at the individual. For example, a road sign such as *Caution sharp bend* or *Caution slip hazard* are secondary aversives because drivers who ignore them may cause an accident.

An aversive, in itself, is not a punisher; it only becomes a punisher when it is used

This road sign is an example of a secondary aversive.

to suppress, weaken or stop a behaviour. Let's take a look at the following example of Capuchin monkeys who were kept in an open, free-running zoo in the Netherlands:

The small Capuchin monkeys started causing problems with the visitors, stealing smartphones, sunglasses and wallets from them. Worse still, some of the monkeys started jumping on to pushchairs and grabbing sweets out of the hands of children. Looking into the problem and using the *think, plan, do* method, I outlined the principles of operant conditioning to the staff and taught them to use spray bottles with accuracy and precision timing. We found that when a monkey was getting too close to a pushchair, a well-aimed spray of water halted the behaviour. Within just a few days, there were no further pushchair incidents! Of course, a member of staff was present, and the spray bottle, attached to their belt, was highly visible. Now the spray bottle, in combination with a staff member, had become a secondary aversive.

Initially, the water spray is defined as an aversive, but when it results in the monkeys stopping their behaviour, i.e., no longer jumping on pushchairs, it could be redefined as a punisher. However, if it is a warm day, and the monkeys appreciate the water spray, they would continue jumping on pushchairs - the water spray would not affect their behaviour. In this situation, the water spray is not an aversive, and, equally, it is not an effective punisher.

It is not the tool, in itself, that can be categorised as an aversive or a punisher - it is the effect it has on behaviour.

There is a tendency among trainers to categorise training tools into what is allowed, and what is forbidden. However, I think we should take a closer look at what is actually going on and work out the effect on the (unwanted) behaviour. For example, practitioners involved in carrying out uncomfortable/painful procedures often use high value food as a distraction from the distress the dog is experiencing. Of course, this is a good approach but, if it is used poorly, it can come back to bite. The dog may associate the high value food with the painful procedure that is taking place. If this happens, the food will be 'poisoned' and will be no further use in training.

Primary and secondary aversives are types of consequences that weaken or suppress a behaviour.

There are a few rules to keep in mind:

- The aversive used must be strong enough to work, i.e., to stop the behaviour.
- An aversive applied *after* the unwanted behaviour has stopped will not work. To be effective it has to happen while the behaviour is taking place or, even better, just before it starts - at the trigger point.
- The aversive must be presented in such a way that the animal does not associate it with the trainer, or the fear/avoidance will be transferred to the trainer.
- An aversive should always be the very last resort, where there is no other option available. For example, to prevent a serious accident, or to save a life.

THE SIDE-EFFECTS

In order to change an (unwanted) behaviour it may be necessary to use some form of aversive. However, there are some very tangible side-effects:

- If the aversive is applied by the trainer, it can cause problems because the animal is likely to become frightened of the trainer. This fear is then easily generalised to the training situation, the location, the building, people, objects, clothing, smells, sounds and much more.
- The use of aversives can cause the animal to show aggression towards the trainer or others in the vicinity.
- Animals may imitate this style of punishment and direct it towards other animals or humans. It's like children who grow up in violent families; there is a high risk that they will later become violent themselves.
- When aversives are used to get rid of unwanted behaviour, suppression only weakens it. The targeted behaviour has not gone away, it's still there and will come back unexpectedly.

Avoid using aversives whenever possible - they often have nasty consequences for both animal and trainer!

NEGATIVE PUNISHMENT

When a trainer takes away a positive reinforcer, we describe it as negative punishment. For example:

- A dog tries to steal food from the hand, so we remove access to the hand.

- A child hits his brother on the head with a toy car, so the mother then takes the car away from him.
- A dolphin trainer appears at the edge of the pool with a bucket of fish. He signals the dolphins to come and when they don't respond, he walks away without feeding them.

A 'time-out' - halting an 'unwanted' behaviour - is also a form of negative punishment. It has a milder effect than positive punishment and has a different emotional impact. This is a form of 'punishment' that is used by many trainers who claim their methods are 100 per cent positive.

A trainer should never use positive punishment as a means of achieving a training goal. Remember, punishment is used as a means of getting rid of behaviour but, more often than not, it just suppresses it. Punishment will also result in the animal offering fewer new behaviours. The general effect of punishment is to suppress all kinds of behaviour, which should be taken into consideration when making training decisions. For more information *see Chapter 17: We need to talk about punishment*.

GENERALISATION

This is a powerful technique that is often not well understood by trainers. There are two types of generalisation - stimulus generalisation and behaviour generalisation:

STIMULUS GENERALISATION

An animal can learn to respond to a specific stimulus. A stimulus that is similar to the specific stimulus can also trigger the same reaction. A dog who has been trained to pick up an item at the point of a finger may also pick up items that are indicated with the whole hand. A dog who has been trained to come back to his trainer on the whistle may also come back if the trainer suddenly blows a different whistle. In nature, two stimuli are never exactly the same, so an animal must respond to a range of stimuli rather than a single stimulus.

Our radio-controlled directional camera dogs receive their directional cues such *as "left", "right", "forward" or "back"* via a human voice over the radio. We use many different radios in training so that we can swap them over in the field if there is a technical failure. In contrast, when training our scent detection dogs, we make sure that they only react to a very specific odour, and not to those that are similar.

BEHAVIOUR GENERALISATION

An animal can learn to exhibit either a specific behaviour, or a group of behaviours, in response to a specific stimulus. For example, we train a dog to hold an item in his mouth, walk 50m and drop it into a specific box. After a few correct repetitions, the dog may start doing some other things as well. He may drop the item several times, push it toward the box with his front paw and then throw it into the box. The result is the same, the item is in the box, but in a different way than we intended. The consequence of the dog's behaviour is the same, but the approach is different. The dog has generalised the procedure of carrying the item to the box and dropping it in.

PROBLEM-SOLVING

An animal that generalises will be better at solving problems. We can see the difference as we work towards a training goal, from one situation to another, and also how different stimuli can affect problem-solving behaviour.

Stimuli can work for us, or against us. Dogs are opportunistic animals and may take short-cuts to get their reinforcement. We often see this when training scent detections dogs. We want the dog to use his nose to find the target scent, but it is easier, and uses less energy if he uses his eyes. For example, when working in an airport, the trainer may hide the target odour in a suitcase. Within several repetitions the dog is relying on what he has seen and is alerting on suitcases - giving false indications - when the target scent is hidden in a cardboard box or behind a fire extinguisher.

Some animals are really good at generalising behaviour. If we train a behaviour particularly well, it becomes easier to train another, similar behaviour. For example, if we are training a dog to complete an agility course, we would start by training the easy obstacles and then move on to those that are more challenging. If a dog is being taught to retrieve, we may start with something easy, like a soft toy, and progress to retrieving a dumbell. Yes, of course this also has a lot to do with the reinforcement rate: simple tasks/ behaviours are shown correctly more often, rewarded more often, so have a higher success rate which, in turn, strengthen the dog's confidence, encouraging him to try new things.

But problem-solving behaviour can go even further. Scientists have tested this by putting animals in a maze to see how they can find their way out. They have also experimented with problem solving tasks. For example, the animal has to figure out how to open a flap or operate a lever in order to get food. Thorndike was one of the first scientists who used graphs and time scales to measure the behaviour of animals, primarily using cats and specially constructed puzzle boxes.

Each cat was placed in a small wooden box and had to work out the puzzle - pulling a rope or operating a lever - to open the door of the box and find the food placed outside it. The door of the box was barred so the cat could smell the food and Thorndike could see inside and record every attempt that the cat took to escape. After monitoring the behaviour, he noted the time it took for the cat to escape and then repeated the procedure. After conducting a series of tests, he established that cats would escape quicker, with less behaviours, i.e., attempts to escape, when they become more experienced. Therefore, he proved that behaviour changes as a result of its consequences.

We also know of birds with amazing problem-solving skills, such as crows using twigs as tools to get food. The more successful an animal is in solving initially simple, and then increasingly difficult, tasks, the greater his motivation - and also his ability and creativity - to solve complex problems. It learns to learn!

Chapter 10:
TRAINING TOOLS

We discovered in the previous chapter that the five key elements – stimulus, reinforcement, extinction, aversives and generalisation – provide the road map for training animals. However, I am writing this book for dedicated, enthusiastic trainers who want to be challenged, who want to take an interest in the science of training, who want to learn more and stay motivated. So, in addition to the five key elements, there are many other tools we can add to the training toolbox which will be the means of building a professional training programme and will help to change behaviour more quickly.

During my time as a coach, I have worked with some with very smart and talented university professors who understood, and could explain, everything about animal training, but were unable to transfer their knowledge from their heads to their hands. Even clicking a clicker was

TRAINING WITHOUT ANIMALS

To help my students practise their manual training skills, I have devised some games to play – without involving animals.

- The 'tennis ball/clicker game' involves observation, hand-eye coordination, and the use of a bridging signal, in this case, a clicker.

This is how to play:

- The student holds a clicker while the coach throws a tennis ball against a wall.
- As soon as the ball hits the wall, the student clicks. After a few repetitions, the sound of the ball hitting the wall and the clicker should occur simultaneously.
- We can now make it a little harder and click when the ball hits the ground after first hitting the wall.
- The 'delivery of reinforcement game' is the perfect way to practise giving rewards, testing both speed and accuracy. To play, we need a few large cups or small buckets, which are numbered, and lined up on a table or in front of a wall.
- The student takes a shallow bowl of kibble, steps back, and throws a piece of kibble into one of the target containers – as if it were the mouth of an animal they were training.
- When the student is achieving a high success rate, we can make it harder. Now the coach shouts out the number of a container and the student has to throw the kibble into the correct container as quickly as possible. Even as they are doing this, the coach is shouting out one new number after another so the game gets faster and faster.

difficult for them. Training animals is also a skill that needs to be acquired. We need to be aware of our body language and be able to control it. The slightest movement of the shoulders, the head or the hand can tell the animal that reinforcement is coming. And at that moment – instead of completing the behaviour – he will stop and look at the trainer in anticipation of a reward.

We need to be able to bridge the signal with a clicker, whistle, voice, or electronic device but, beyond that, we need to be aware of how to present and give the reinforcer, the impact of our body position, and how we move around in the training environment.

I am all about plans, protocols, data and science. But for me, training animals is not just a question of working with intellect, manual skills, or food and ball machines. I believe we should also use our hearts and our gut instinct, as this can act as a powerful social reinforcer. The scientific approach is to focus on the evidence, finding the sequence of tasks or behaviours that produce the same result from the same procedure. However, there is more to training than working with stimulus and response.

For example, if I give my dog a cue to sit, and reinforce him with something he really likes, it does not guarantee that I will get a perfect sit the next time I give my cue. We train with our hearts as well as our heads; it's about our emotional response, our timing, and our gut instinct. If I am not particularly happy with my dog's execution of the sit, I may use my clicker and throw him a treat but this reinforcement is, at best, half-hearted, I am not investing emotional energy in my response. Imagine I am engaged in search and rescue and, in training, I have reinforced my dog with his toy hundreds of time when he makes a find. Now imagine I am on a real-life operation, and my dog locates a survivor – now my reinforcement is off the scale! It's not about my dog's favourite toy, it's my huge emotional response that makes all the difference. This type of social reinforcer is truly powerful!

So yes, we should also follow our gut instinct when we train. We just need to do it – to trust our sense of intuition, just as we would when coaching people or raising children.

OPENING UP COMMUNICATION CHANNELS

When training animals, we do not share a spoken language, but this does not mean we have to complicate the pathways of communication. My advice is: keep it clear and keep it simple. All we need to do is explain to the dog what we want – and convince him that it is what he most wants to do. For example, if I want to train a dog to stand on a platform, it will really help if he has a strong desire to do it.

Dogs are our companions and there are times when we talk to them as though they were human. We may chat to them on a walk, or we may confide our feelings if we are worried about something. However, we know from scientific research that they don't understand language in the way we do. They don't listen to all the words in a sentence and, even if they had a full understanding of what we were saying, they cannot answer back. There is nothing to be gained in asking a dog why he peed in the living room while you were out or explaining why you disapprove of something he is doing. Dogs cannot understand language in this form - but they can understand words. And it's not just to do with sounds, as some people would like us to believe. As evidence I cite the example of Chaser, the Border Collie who was trained to identify over 1000 toys by name. When her guardian/trainer, Professor John W. Piley, named a particular toy, Chaser went to look for it, identified it, picked it up and retrieved it.

In our work with radio-controlled camera dogs, we discovered that dogs generally have better communication skills than we think. A dog will observe his trainer when given a verbal cue, paying more attention to the body language than the spoken word. A nod of the head will be his cue to lie down, and a slight raising of the head will be his cue to sit. We tested this by swapping our verbal commands *"sit" and "down"* for *"car" and "apple"* and an estimated 95 per cent of dogs continued to sit or lie down based solely on body language.

This is not the case for our camera dogs, who are remotely controlled by radio. They operate out of *our* field of vision; they cannot see us when they are inside a building and we are outside, or when they are working at a distance of 300m in pitch darkness. They must rely on the verbal cues that are transmitted through a radio on their collar. And yet they understand the difference between *"left", "right", "forward', "here", "sit", "down", "look", "get", "let go", "touch," "hide"*, and more. It's very impressive!

So yes, I believe dogs are better at communicating than most trainers realise, but our communication should still be kept as simple as possible. In training, we often give unintended signals through the way we move, making sounds in addition to our verbal cues, plus signals that come from the environment. We need to pay close attention to exactly what we want to train and omit all other communication.

When communication problems arise, it often means that we have progressed too quickly, and the dog doesn't understand what we want him to do. If the form of training we are using has worked well in the past, it could be that our technique has changed a little, and our 'message' has lost clarity. In this situation, it is helpful to train incrementally, breaking up the behaviour into smaller chunks. We should also take a closer look at the training environment. Is the dog really focused on the trainer? Are there distractions we have not taken into consideration? For example, there could be a scent, such as the odour of a bitch in season, which we cannot perceive, but would be a major distraction for the dog. Or there could be a high frequency noise that we cannot hear but is driving the dog insane. Remember, in any training situation, the dog must be able to give his complete focus to the trainer. In this way, he will start to learn that the signals from the trainer are linked to the behaviour he should perform.

ABC OF TRAINING

Just like the communication element, I like to keep overall training simple, and there is no better starting place than the ABC method, developed by behaviourist, B. F Skinner. It works on the following, straightforward principles:

A = Antecedent: This is everything that precedes the behaviour. We focus here on the cues or triggers that produce the behaviour.

For example, a dog is given the cue to sit.
B = Behaviour: This is the behaviour that is happening right now. Remember the 'dead man test' when describing and analysing behaviour? (see *page 47*).

For example, the dog settles into a perfect sit.
C = Consequences: They have a direct effect on the behaviour and the learning curve. Positive consequences lead to more of a behaviour, making it stronger, faster, and longer lasting. Bad consequences lead to a weaker behaviour, and less of it.

In this positive example, the dog is rewarded with a tasty treat which will ensure, going forward, he will produce better and faster sits of longer duration, in response to his 'sit' cue.

Once we view behaviour through the lens of the ABC method, everything becomes easier. It ties in with the well-known saying:
If you can't explain it simply, you don't understand it well enough.

Even the most complex behaviour can be broken down into smaller steps if we:
- Write down what we see.
- Film the session using slow motion footage to break down everything down into smaller segments.
- Apply the ABC of training.

This objective assessment is key to understanding exactly what is going on in the training process.

SETTING CRITERIA

One of the most important elements in a training session is deciding on the

UNDERSTANDING ANTECEDENTS

Put simply, antecedents are anything that precedes behaviour. It is the cues or stimuli that can increase, but also decrease, the likelihood that a certain behaviour will be exhibited. It may come from a single source, or it may be a combination of several signals or stimuli.

For example, for a search and rescue dog, the sight of a collapsed building can be one of the antecedents to start searching for people under the rubble. Most often it is a combination of:

1. The sight of the building.
2. A verbal signal from the handler.
3. Special clothing worn by the handler, and equipment that is used when searching.

If any of the three things are missing, it can affect the dog's behaviour:

- If antecedents involved in triggering behaviour lead to positive reinforcement, the consequences are pleasurable. This will affect antecedents in the future – making them stronger
- If the behaviour has negative consequences, an antecedent can also become negative for the dog, meaning he will be reluctant to engage in the behaviour and may even run away.

criteria and sticking to it. If criteria are not established, there is no point in trying to train. Of course, we could train just for fun or for physical and mental exercise, but if we want to make progress, we need to set criteria. Setting critera allows us to:

- Measure whether there have been improvements in our training.
- Decide if we need to train more at this level, whether we need to make it easier, or whether we are ready to progress.

Establishing criteria is not complicated – think of it as a short video recording the behaviour we would like to achieve.

Let's look at the criterion we might set for a 'sit' in obedience training. This could be: *the dog sits within two seconds of the signal and stays seated for five seconds*. Therefore, if the dog fails to sit within two seconds, the exercise is halted and there is no reinforcer. If the dog sits within two seconds but gets up again before five seconds have elapsed, again, the exercise is halted without a reinforcer.

Criteria creates clarity in both boundaries and rules. Imagine a football pitch with no marked lines, and a match played without rules. Where is the fun? The 'match' has no meaning, and there is no opportunity to improve or progress. If we want to grow, reach the next level, and measure quality, we need to apply criteria to our training.

HOLISTIC LEARNING

We may apply rules to our training, but we need to understand that training is not a linear process. It does not consist of the 'sit', the duration and the reinforcer. Rather, everything is interconnected; one element of a behaviour will affect the next part, and this will impact on the wider context of the whole training programme.

Imagine filling a glass of water. If we were to think of it as a linear process, we would:

- Turn on the tap.
- Hold the glass under it.
- Turn off the tap.

But it's not like that; the process is more like a circle. We need to consider:

- The position of the glass – ensuring it is not held too high or too low from the tap.
- Turning on the tap – turning it on a little will result in a trickle of water, turning it all the way will result in a jet of water.
- The level of water in the glass – how much water do we want in the glass?

It's all about cause, effect and result.

Training a dog does not have a start and an end point. A good and effective animal trainer needs to think in terms of

Filling a water glass is not a linear process.

systems, where each step affects the next. If we change one part of the system, it can affect other parts, or even the system as a whole. It's not just about giving a reward; it's about how and where it is given. For example, let's say we are teaching a dog to walk on the right side of a wheelchair. If he runs too far forward, we don't jerk the leash. Instead, we give the reward a little further back so that he learns the correct place to position himself.

Returning to the example of training a 'sit' in obedience, we want to see behaviour within two seconds of the verbal cue. But, at the same time, we want the dog to 'sit square' with his front legs in line with his hindquarters, we want him to focus on the trainer, without barking or whimpering and maintain a relaxed, open mouth. We want to see all of this for at least five seconds in this training phase before we give a bridge signal and let the reinforcer follow. From this we can see that there are many parts of the system that require the trainer's attention, analysis, and decision-making, before proceeding to the next step.

A trainer's job is to make decisions!

- Do we strengthen: yes or no?
- Can we help the dog?
- Are we ready for the next step?
- Do we need to practise the behaviour for longer?
- Do we reinforce with food or with play?
- Do we continue training, or do we take a break?

These, and many other decisions fall within the trainer's remit, and most of them need to be taken in a short space of time. That is why I like the *think-plan-do* principle so much (see *page 17*). If we take the time to think and plan, we can make quick decisions when it matters most. Many trainers are tempted to go straight to the *do* phase but if they are unable to make the necessary decisions, their training will slow down or become problematic. So, choose to take the time to think and plan first!

BRIDGE SIGNALS

A bridge signal, often known as a marker signal, provides the animal with a clear sign that a reward is on its way. It doesn't matter what type of bridge signal is used, as long as it's clear and the animal can hear it. The most common example of a bridge signal is a clicker, but it can also be

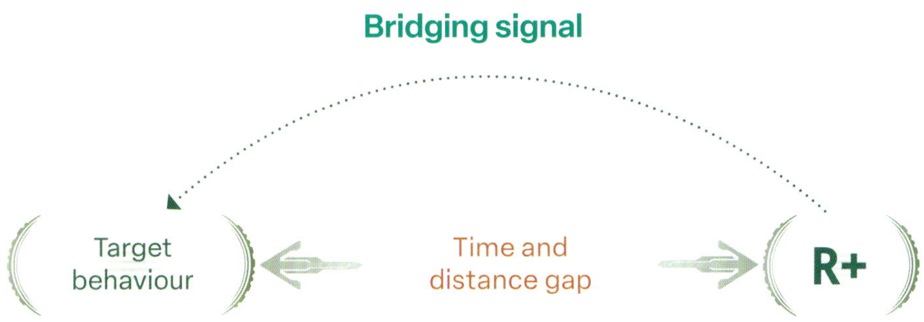

The bridging signal closes the time and distance gap between the target behaviour and the primary reinforcer

a whistle, a light, a hand signal, a word, an ultrasonic sound, a touch, or anything else that the animal can readily distinguish.

I prefer the term 'bridging signal' to 'marker signal' because it implies bridging the temporal and spatial gap between the target behaviour and the reinforcer. In many situations, the trainer is close to the dog and can throw the reward quickly. But as the training progresses, the time lag and distance increase. Now the bridge signal will play an important role in keeping the communication clear. And it's a very precise tool for capturing the exact moment of behaviour we want to reinforce!

MOTIVATIONAL TRAINING

As already highlighted, motivation plays a central role in animal training. An unmotivated animal will not participate. Most dogs from working breeds or lines have something called intrinsic motivation; it's in their nature, in their DNA. For example, when Pointer puppies encounter a bird for the first time, they freeze. When Retrievers see something fall, they run and pick it up. Border Collies instinctively herd sheep. Malinois like to dig their teeth into a biting arm. German Shepherds love tracking.

In addition to intrinsic motivation, there is external motivation, such as play, food, activity, etc., which we use as our reinforcers. The better we are at motivating our animals, the more they will want to work for us and to engage in our 'games'. Because, at the end of the day, that's what we ask of them: to play our games in return for the external motivation we provide. Dogs are, without doubt, the easiest of all animals to train because they are already so focused on us.

Motivation: Charlie can hardly wait for the day's work to start!

To build motivation, we need to find the primary reinforcers, i.e., those which are most valuable for that particular dog, such as food or play, and link them to a secondary reinforcer, such as a clicker. Following a series of fast click-food repetitions, a dog who loves food (primary reinforcer) will quickly learn that the sound of a clicker (secondary reinforcer) predicts that food is coming.

When I am working with a dog I use several reinforcers, each of which has which has a specific value for him. For example, I might rank my reinforcers as follows:

- Roast chicken
- Cheese
- Kong
- Ball
- Frisbee

This allows me to use motivators most effectively to influence the outcome of training. Remember, we always get what we reinforce! *For more information on grading rewards, see Chart toppers, page 115.*

Rewards can be graded: small achievements warrant low-key rewards; big achievement deserve jackpot rewards (see *page 114).* But all in all, rewards will motivate the dog. The often heard: "*He should do it just for me*" is, in most cases, wishful thinking. As mentioned earlier, we need to ask ourselves what is worthwhile about a certain behaviour for the dog (see *'what's in for me?', page 49*). The owner's satisfaction – evidenced in the absence of sanctions – is usually insufficient motivation for a dog, even if there is the rare individual who will go to great efforts to please his people.

GET THE TIMING RIGHT!

I have already talked about the importance of decision-making, bridge signals and reinforcers. However, these elements will only be effective if we get the timing right. I learnt this lesson, and acquired the necessary skills, when I was attending the chicken workshops run by Bob and Marian Bailey. One of the basic skills is training a chicken to peck right in the middle of a target. At the precise moment the beak hits the centre of the target, the trainer has to activate the clicker (bridge signal) and give food (reinforcer). If the trainer presses the clicker too early, the chicken will not peck at the target, but will hover with its beak a few centimetres above it and wait for the click. If the click is too late, the chicken will peck several times in a row instead of just once. While a dog's response is slower than a chicken's, timing is still important.

We also need to be aware that a new signal must always come before one that is already known. Take the example of a dog who responds to a hand signal to sit, and we now want to train this behaviour on a verbal cue. We, therefore, start by giving the verbal cue, *"sit"*, followed by the familiar hand signal. Attempting the sequence the other way round – first the familiar hand signal and then the new verbal cue – will not work. We humans can understand this sequence, but it is too difficult for dogs.

If we want to introduce a new signal for a known behaviour, the new signal must always come before the old, known signal.

The sequence a dog understands is anticipation of cause and effect. If the 'sit' behaviour is triggered by the hand signal, it will result in a reward. Anything after the hand signal has no meaning for the dog. But when a verbal cue, *"sit"*, is followed by a familiar hand signal, the new verbal cue is of interest. Now the dog can anticipate that the hand signal will follow. After a few repetitions – and a slow phasing out of the hand signal – the dog will sit solely in response to the verbal cue.

Getting the timing right is vital if training is to be effective. However, the best coach in the world will still make mistakes in timing and judgment. We can get round this by using electronics to introduce absolute precision to our timing. For example, we have developed a high-tech scent wheel which can be programmed to reward the dog when his nose has been on the 'hot' target for the desired duration. In addition, each target gives us electronic feedback, in the form of data, recording the length of time the dog's nose is in position.

We have developed our own high-tech scent wheel, with infra-red sensors, which helps us to be precise with our criteria.

It is worth noting that our inability to be absolutely precise and consistent in our timing is a strong argument against the use of positive punishment. If this form of punishment is to lead to the desired result, it must be strong enough, and take place at exactly the right moment, otherwise the situation will be made worse. For example, imagine a dog is trying to bite another dog and the guardian responds by pulling back on the leash and yelling *"no!"*. However, the guardian is not sufficiently strong to stop the dog lunging forward, and his verbal reprimands are ignored. The dog gets close enough to bite. The fight breaks up and dog and guardian move off in the opposite direction. What will happen when these two dogs meet the following day? Yes, the fighting will restart, probably with increased intensity.

In contrast, if our timing in not spot-on when using a positive reinforcer, the worst-case scenario is not getting the desired behaviour (or not getting it so quickly), but at least our relationship with the dog and our mutual sense of trust has not been damaged.

If you don't have good timing when using positive reinforcement with food, you may end up with an overweight dog.
If you don't have good timing when using positive punishment, you will ruin your relationship with the dog

TRIGGERS

Triggers, like antecedents, precede behaviour, but unlike them, they not only increase or decrease readiness for a particular behaviour. They are the last, crucial stimulus that triggers the behaviour, like flipping a switch.

All triggers are antecedents, but not all antecedents are triggers.

Often the term 'trigger' is used with a negative connotation:

- One dog triggers another to be aggressive.
- A running rabbit triggers a dog to chase near a busy road.
- The smell of a decomposing animal carcass triggers a dog to roll.
- A visitor at the front door triggers a dog to bark.

Triggers are all around us and, depending on the individual, specific triggers prompt specific behaviours. If we can identify these triggers – great! We will learn more about the dog, the environment, the context, the effectiveness of his training, and the strength of the trigger. This means that we can use triggers to our advantage.

If the trigger starts the desired behaviour, then we know what to do. We either put a signal in front of it or start the reinforcement process right away. But when a trigger prompts a negative behaviour, it is often the case that trainers no longer know what to do. We spend most of the time trying to avoid the trigger that starts the unwanted behaviour. If we meet another dog, we cross the street, or we avoid the field with the rabbits. We 'warn' the dog, telling him not to respond to the trigger that is likely to come. We say *"no"* or keep the leash tight and short as a precaution. Now it is important to understand: these actions also become antecedents and the dog will anticipate the behaviour that stems from them. He learns*: Rabbits are coming soon and I'm going hunting!*

However, we can turn the triggers of negative behaviour into something good, i.e., we can train our dogs to turn the triggers into something that is good for them. Remember the '*what's in it for me*' principle (see *page 49*). For example, when dealing with a reactive dog, we can use the 'other dog in sight' trigger to make him look at us and reinforce the moment of eye contact. The trigger of a rabbit coming into view becomes a signal for the dog to sit down quickly and then get an extra special reward.

Avoiding triggers that lead to negative behaviour removes the opportunity for this type of training. However, once we understand that we can use triggers to turn a negative into a positive, there are no limits to our training. We need to observe, collect data, analyse the triggers, understand and consider the context – and make use of these findings in our training.

Make use of triggers that elicit unwanted behaviour instead of just avoiding them!

LUMPING AND SPLITTING

Lumping is the term used when we simply see what behaviour a dog will offer in a particular situation, in relation to the environment, or if he is given a problem

ACTING ON SUPERSTITION

Often, without our realising it, dogs will engage in behaviour that they think will result in a reward. For example, when the dog is asked to sit, he responds, raises his right front paw and is reinforced. From his perspective, reinforcement follows not after the sit, but after he sits and raises his right front paw. Obviously, he will keep on repeating this as it is the key to getting reinforcement. Often dogs think they need to give a series or sequence of little behaviours that will lead to reinforcement, whereas we humans focus on the primary, most visual behaviour.

Incidentally, this also applies to human behaviour. For example, imagine a soccer player who is about to play the most important game of the season. In the changing room he realises that he has the wrong socks: instead of two blue socks, he has one red and one blue. It's too late correct the mistake so he plays the game wearing different coloured socks. He scores three goals – it's the best he's ever played. What colour socks will he wear for the next game? A red one and a blue one, of course, because he now believes they will work for him and make him lucky!

to solve. In training terms, we often apply it when a dog is new to us, and we start training without set criteria. We might try some random cues, and maybe use body language to communicate. We might throw a ball, introduce some food, or start a game. We want to see how the dog responds and get an overview of the behaviours he might offer.

The advantage of lumping is that we cover a lot of ground and see a lot of different behaviours in a short amount of time. However, there is a strong likelihood that the dog will do something contrary to our requirements. For example, he may hold an article the wrong way, or use an unconventional posture for an alert. So, in order to train the required behaviour, we would need to interfere.

In contrast, we can split or break down a behaviour and work on it, step-by-step, according to the ABC formula (see *page 86*). There is no doubt that this is time consuming, but it does lead to strong, reliable and well-conditioned behaviour. Let's take the retrieve as an example:

- We start with asking for a sit.
- We work on the dog waiting for his cue and not moving when we throw the dummy.
- We encourage the dog to run out and start searching.
- We train the pick-up and the return.
- We work separately on how the dog holds the dummy and on the speed of his recall.

SHAPING

The process of shaping is also known 'successive approximation' or 'differential reinforcement'. B. F. Skinner, and his students, took the lead is showing us how powerful shaping can be on many different

A chicken is trained by shaping to peck in the middle of a target.

animals. Following on from his work, Keller and Marian Breland presented the shaping technique to the world at their *IQ Zoo*.

Shaping involves giving or withholding reinforcers to change an animal's (or a human's) behaviour. Let's go back to the example of training a chicken to peck right in the middle of a given target. In this scenario, a round target is placed on an empty table in front of the waiting chicken:

- As soon as the chicken looks at the target, the trainer clicks and gives a food reward.
- After a few repetitions the chicken no longer gets a reward for looking at the target. Instead, the trainer waits until it makes a slight movement towards the target before clicking and rewarding. This has the effect of increasing the frequency of pecking towards the target.
- Now the trainer 'clicks' the chicken as it gets progressively closer to the target, until it pecks the target itself.
- If the chicken does not peck in the middle of the target, but is a little to the left, the trainer clicks, but then gives the food reward a little to the right of the target.

This example shows that shaping is a

combination of using a bridge signal and feeding.

Let's take a look at training a dog sitting on a chair. In this scenario, the dog already knows that the sound of the clicker announces a food reward.

- A chair is placed in a small space with no distractions. There is no other furniture in the room and nothing hanging on the wall. Only the chair, dog and trainer are present.
- When the dog looks at the chair, the trainer clicks and rewards.
- This is repeated a few times until the dog begins to show increased interest in the chair.
- The trainer now 'asks' for more behaviour: instead of rewarding the dog for looking at the chair, the dog must take a step towards the chair to earn his click and reward.

Note: If we expect too much at once and take too big a step, the reinforcement rate will become too low, meaning we reward less often and the dog will no longer be motivated to work for us. He loses interest in the chair and the training. Conversely, if we take small steps, reward often, and ask for just a little bit more, we can 'shape' the dog toward the chair. I recommend working in short, two-minute training sessions, or 5-10 clicks. After that, take a little break before starting again. After about 10 sessions, about 40-50 clicks and rewards, the dog will understand he must have two more paws on the chair – as training progresses, he will understand that he must sit on the chair.

This process involves a series of successive approximations. In this way, we can turn an 'unlikely' behaviour (the dog walking up to a chair, putting his paws on it and then sitting on it) into a very 'likely' behaviour in a very short space of time. We are building a new behaviour out of previously undifferentiated behaviour, i.e., a behaviour we have not tried to manipulate or influence. Continuous reinforcement is given every time the dog touches the chair with his paw, progressing to differential reinforcement, i.e., only reinforcing the dog when his paw is on the chair – not under it, or to the side of it.

A chair in an empty room is the target for shaping a behaviour.

Think of the 'hot or cold' game that children play. One child is sent out of the room while the others hide an item. The child is then brought back into the room and is tasked with finding the hidden item. The only guidance is given by the other children, shouting *"hot"* when the 'seeker' is getting close to the hidden item, and *"cold"* when they move further away. A simple matter of shaping.

Shaping is an important process that every trainer must master. Following on from Bob and Marian Bailey's chicken workshops, I also use chickens to help my students improve their shaping skills. Chickens are very fast-moving, and they also have very sharp eyesight, which means that they are merciless if a trainer makes a mistake. So, when training a chicken to peck the middle of a target, the chicken notes that the trainer's thumb tenses ready to press the clicker (the bridge signal) which signals the reward. Within two or three repetitions, the chicken does not wait for the click; the muscle tension has become the bridge signal! As soon as the chicken notes the increase in tension, it runs towards the trainer for food (reinforcement) without actually touching the target. This illustrates the importance of timing – and the importance of remaining as still as possible so there are no unintentional clues to muddy the waters.

LURING

Luring involves 'bribing' an animal to perform a certain behaviour by guiding/luring the behaviour with food or with a toy. It's a quick and powerful technique, but it is also a risky one. If we lure for too long, the animal loses interest in offering the behaviour and makes less effort. He will only perform that one behaviour, in that specific context and then, only then, when the reinforcer is present. As training progresses, it becomes difficult, if not impossible, to demand more or different behaviours.

Let's take the example of the dog sitting on the chair. With shaping it may take us 10 training sessions, with 40-50 reinforcements in each, until we get the dog where we want him – sitting on the chair. If we were using luring, we would put a food reward on the chair, and we would probably get the dog on the chair in just one training session, with just 3-5 reinforcements. But then we must quickly start the process of shaping before the dog loses interest or contextualises the behaviour to the extent that that he will only perform it in one given set-up.

I like to work with luring because it is quick and powerful and saves a lot of time. But we need to be aware that luring is just the beginning of a training process before we progress to shaping. We must, therefore, condition the dog to a bridging signal, such as the clicker or a whistle, prior to luring. The dog should also have some experience of shaping, as this is a required basic skill if training is to progress.

The big difference between shaping and luring is the reinforcement history. Let's take 'sitting on a chair' as an example. In the process of shaping a behaviour, we reinforce approximations of what we want to guide the dog towards our intended goal. Initially we use a high rate of reinforcement because it's not easy to get the dog's attention to the target area (the chair) and to perform the intended manoeuvre (sitting on the chair). These multiple reinforcements, given when the

If you shape the dog on to the chair as a target, you will have many reinforcements away from the chair. When luring, he will be reinforced directly on to the chair.

dog is in the vicinity of the chair, create a reinforcement history, which can become problematical. The dog has learnt that reinforcement comes when he is anywhere near the chair, but he is not aware that the chair is our goal. In contrast, when we use luring, all reinforcements take place directly at the chair, making the target area crystal clear. The dog is more likely to perform the wanted behaviour because he has already figured out that the reward must have something to do with the chair.

TOSSING FOOD

This isn't exactly a training tool, but it is a skill that every trainer should acquire – I work with it all the time.

Food rewards give us the ability to reinforce a lot in a short amount of time, and in almost any situation. This is enhanced if we use food treats in a targeted manner. Take the example of training the 'sit' for obedience. We get best responses when we reward quickly, and in the desired position. A bridging signal, such as the clicker, tells the dog: *"Well done, the reward is on its way"*. But tossing the food for the dog to catch is also a powerful reinforcer, making the behaviour stronger because good things (food) keep coming.

Importantly, the dog must be able to catch the treats so we must aim directly for his mouth. If he gives a perfect sit – and the treat lands on the ground – he will have to abandon his sitting position to get to it, so we will be inadvertently reinforcing him for breaking the sit. We, therefore, need to practise our technique before we start training. This is what to do:

- Line up a row of small buckets and number them. The trainer stands at a distance of 5m, equipped with food treats.
- The trainer's task is to throw a treat into each of the buckets as quickly as possible. Initially, progress is likely to be slow, with lots of 'misses', but practice makes perfect!
- When the trainer is achieving a success rate of 80 per cent, or more, we are ready to introduce the dog. I like to use a small platform, stationing the dog in a sit, and the trainer at a distance of 1-2m.
- The dog needs to be focused on the

trainer, as he needs to learn the body language that is integral with throwing treats. For this reason, the trainer needs be consistent in their movements, getting the treat and then aiming for the dog's mouth. No verbal cues are required.

Marieke deliberately throws pieces of food while Ygor is sitting on the platform.

- If the dog misses the treat and it falls to the ground, he must not be allowed to leave the platform. Instead, the trainer retrieves the treat as quickly as possible, and restarts the exercise.
- With practice, the trainer's aim – and the dog's ability to catch – will improve. At this stage, the trainer can speed things up and, if this works, start throwing from different positions, using different body language, and at different speeds. This creates the opportunity to reward, at speed, from almost any position that is within accurate throwing distance – and the dog will love it!

CHAINING

Breaking down a behaviour, and training the elements individually, facilitates learning and strengthens understanding. But once the individual steps have been trained, they need to be linked to form a chain, i.e., the finished behaviour.

Let's take retrieving a ball as an example. To achieve the end behaviour we need to condition the following behaviours as individual steps:

- Sit at the start.
- Search in the field.
- Find the ball.
- Pick up and hold the ball.
- Carry the ball.
- Return to the trainer as quickly as possible.

All of these behaviours are then linked together to form an overall behaviour, known as the retrieve. In this instance, we need to be aware that a behaviour is only as strong as its weakest link. A single weak behaviour can destroy the entire chain. It is, therefore, imperative to condition each and every element, ensuring the dog has a full understanding, before we chain the individual elements together.

With the help of chaining, we can train a dog to perform a wide variety of tasks, such as setting down a robot for remote control via radio.

REHEARSAL

I am a great believer in the power of rehearsal – using repetition as a means of training – because it leads to strong conditioning of the end behaviour. However, we need to ensure that we condition what we want to condition, sticking to our criteria.

I generally build my training plans so that I do approximately 10 repetitions of each training step. However, I must achieve a success of at least 80 per cent or more in this series of repetitions before progressing to the next step in the training plan. This ensures there is no such thing as beginner's luck; the number of reps, plus the success ratio, proves the dog understands the exercise. If we jump from one step to the next, with only two or three repetitions, we will make fast progress, but we will not have established a strongly conditioned behaviour before we start chaining. Remember, it takes many repetitions before a behaviour becomes strong, meets the criteria, and is carried out with confidence and fluency.

STIMULUS CONTROL

For many trainers, stimulus control means being able to keep a dog from chasing a rabbit or engaging in some other form of unwanted behaviour. But, in fact, stimulus control goes much further: it is the essence of animal training. It is about bringing a behaviour under the control of a stimulus or a command/cue. First the animal must learn *what* to do and then *when* to do it. Take the 'down' as an example:

- The aim is to capture natural behaviour and then reinforce it. So when the dog lies down, the trainer tosses him some treats. As a result, the dog will probably lie down more often to get more treats.
- After a few repetitions, the trainer no longer throws treats when the dog lies down. Instead, the trainer gives a verbal cue, *"down"*, before the behaviour and then throws treats if the dog lies down. The dog quickly understands that lying down, in combination with the command *"down"*, leads to food; lying down without a prior cue, does not. He distinguishes between the audible cue, plus behaviour, which leads to a reward, and all other environmental stimuli which have nothing to do with this sequence of events. He must also learn that *"sit"* or *"here"* does not mean lie down.
- Once the dog has assimilated this information, we want him to generalise

TAKE A BREAK

It's always beneficial to give the dog some time to 'think', i.e., to process the new information he is taking on board during the learning process. Referred to as latent learning, the effect of taking a break – often when we hit a wall in training – can be almost magical, with the dog suddenly being able to perform the desired behaviour next time we ask for it. I don't have scientific evidence to back this up, but it is an experience I have shared with many other trainers. I believe it is hugely advantageous to train for a few minutes and then have a break and start again. This often produces big changes in behaviour in a very positive way.

the new cue, *"down"*, to all environments so he responds not only in the living room, but also in the kitchen, in the garden, out on the street, in the forest or in front of a shop.

From this we can see that stimulus control is the basis of all training. It is about differentiation, generalisation and expectation. We have tested this to the limit with our radio-controlled camera dogs, who must have perfect stimulus control, and distinguish our commands precisely, because they only receive them via a radio on their collar, and out of sight of their handler.

We can test stimulus control by saying words that sound similar to our chosen verbal cue but are slightly different. So instead of *"down"* we could say *"known"* or *"sown"*. If the dog lies down on the wrong cue, we withhold the reward. As well as testing the verbal cue, we can also change our position, for example, instead of standing in front of the dog, we could turn to one side, sit down, or walk behind the dog. We want the behaviour to be so strong that even if we whisper *"down"* while we are lying down and looking the other way, the dog will respond immediately. That is stimulus control!

Behaviourist Ken Ramirez gave me a revealing insight into how animals learn under stimulus control. He was in the middle of a large-scale dolphin show, working through a well-practised routine, when suddenly, the dolphins were reluctant to approach him. As far as Ken and his team of trainers were concerned, it was business as usual; they could not identify a problem. Then one of the trainers noticed that Ken was wearing new shoes. They were virtually identical to his old shoes, but these had white instead of black soles. It was a tiny detail but was enough to affect the dolphins' response.

Protective clothing can make it difficult for the dog to find his handler.

From this we can see that, when training, we should not only vary our body language and tone of voice, but also our dress. This is particularly important for all those who work in uniform, wear protective clothing when on duty, or those who have to switch to working in a team when they have been used to training alone.

FADING CUES AND TARGETS

In a world full of distractions, it is important to make everything as easy as possible in the initial stages of training. I, therefore, design the training environment in such a way that I can control external stimuli. This allows me to practise 'sit', 'down' and 'here' with a lot of body language, clear verbal commands and in a situation in which I am the only focus for the dog.

Our remote-controlled camera dogs start with very large, clearly visible targets, which they use to learn the signals for 'left', 'right' and 'forward'. Herding dogs start training in a small field with a few sheep, gundogs dogs work with dummies, and scent detection dogs with a distinct, target scent. Working over short distances, employing a high reinforcement rate and using clear signals and targets brings clarity to the early stages of learning. As training progresses, we need to slowly taper off these training aids as, at some point, the dog needs to be able to work without them. So, we let our body language become weaker and weaker, we minimise the use of verbal cues, saying them more quietly and using them less frequently, we fade out environmental stimuli, we gradually reduce the size of targets until we don't need them anymore, and we fade out training equipment such as automatic feeders and ball machines. Scent detection dogs learn to search for smaller and smaller objects with weaker and weaker scents, herding dogs work a flock of sheep in a large expanse, and gundogs find the game in a field full of distractions.

When we start to taper off the cues and stimuli, we need to proceed carefully. We know that animals pick up on the tiniest, unintentional hints we give them (see *unintentional influence, page 131*). So we need to make a plan before we start tapering. What is the signal we want or give, or are able to give? What behaviour do we want when we give the signal? What steps will we take to fade the prompts? We need to formulate a plan that guarantees success, and always be ready to take a step back if the dog makes mistakes.

CONTEXT AND GENERALISATION

Dogs are masters at adjusting to context, a skill that is best illustrated by watching them prior to a competition. Take IGP as

When training camera dogs, we gradually fade our body language and the use of targets.

an example. Outside the competition area, the dogs may bark, pull on the leash, and pay no attention to their handlers. But as soon as they step on to the field, it's like a switch has been flipped - the dogs only have eyes for their handlers. No barking, no tugging on the leash; it's all about intense focus and lightning-fast response to commands, a response that is triggered by context - in this case, the competition arena.

A scent detection dog working a line-up, i.e., a series of scent samples placed in a row, will ignore both food and toys that are set up as a distraction. We tested this by videoing one of our scent detection dogs in a trial. We set up eight stations, each with a different odour, such as raw meat, cooked chicken, kibble, a tennis ball, a favourite toy, and other tempting smells. One station was set up with the target scent - a tiny piece of kong, only 1 mm in size. When we played the video, we could see how quickly the dog ignored the 'distraction odours' only to stop abruptly in front of the kong and do a perfect sit alert. That is context. In everyday life, this dog would behave like any dog and his response would be completely different. However, if a dog searches for too long in a double-blind setting, where neither the dog nor handler know whether target scent is present, the dog may become distracted by other enticing smells. Training is an ongoing process. A dog may be able can do a double-blind line-up in training, but that does not mean he is ready for operational work.

Let's consider the example of a dog tasked with searching for narcotics in a house that has just been raided by the police. The people living in the house had just sat down to breakfast and plates with eggs and bacon were still on the table. The kitchen counters were littered with food, the refuse bin was overflowing, the TV was playing, there was a guinea pig in a cage, and the family's pet dog has just been removed by the uniformed police. Speaking from experience, this is when you, as a handler, realise that no matter how proficient your dog has been in training, working to the level of a double-blind line-up, this is something you have never encountered before. The context is completely different. Training must, therefore, replicate real-life scenarios, as far as possible, working outside known environments.

To better understand contextual learning, consider how a dog eats food from the hand. Initially, when I offer food to a young, enthusiastic dog he will simply grab the food and be unaware of whether he is inflicting pain or causing an injury. He just wants the food. I now close my hand, and only open it when the dog is gentle. After a few repetitions, he slows down. He approaches my hand almost in slow motion, using his front teeth very carefully while looking at me. But in a different context - a piece of food falls to the ground, or a kong is offered as reinforcement - that same dog will grab his prize with no inhibitions.

Overall, we can say that dogs are quick to generalise, but we need to teach this by introducing different contexts. As soon a behaviour is established in a particular context, we need to move to the next context and repeat it all over again. After doing this a number of times, the dog is able generalise at speed when placed in new situations. For this reason,

I do not advocate training for a 100 per cent success in new behaviours. Instead, I recommend training for an 80 per cent success rate and then moving to a new context. Both dog and handler need to keep learning together in different contexts so success is not 'connected' to a specific environment.

Some time ago, I was asked by border control to train dogs to find illegal immigrants hidden in trucks, trailers and containers. I suggested using search and rescue dogs, but they were convinced that they needed dogs to be specially trained for the new task, as the search and rescue dogs would not be searching in their known environment, i.e., a context where they had built up experience and gained success. But with some additional training, their skills were transferred to working with trucks, trailers and containers. They simply learnt a new context. Often, it is our way of thinking that is the limiting factor. We need to think outside the box – there is no such thing as problems, only possibilities!

So, yes, dogs can generalise, but we have to help them to do so. A search and rescue dog, trained only on the rubble of collapsed buildings, has no clue what to do in a harbour full of containers waiting for shipment to England. A dog who has been trained exclusively to find people hidden in containers has no clue what to do on a search and rescue mission following an earthquake.

I learnt the importance of contextual learning many years ago when I was working with police patrol dogs. One of the handlers was working with a fully-trained police dog, who had also won some Schutzhund tests. A week after the dog's final examination, they were out on patrol

Expectation is built up before the start of a search.

and were sent to a downtown bar where a man was threatening the occupants with a knife. The handler commanded his dog to attack – and nothing happened. The handler was so disappointed he asked for another police dog, saying his dog was unreliable. But the problem was that this dog had learnt his job in the context of the training ground and the target had always been a 'suspect' in a thick, smelly protective bite-suit. He had never worked in any other environment; therefore, the fault lay with the handler – not the dog.

EXPECTATION

Expectation is a harder concept to grasp and is less easy to measure, but it also plays an important role in animal training. We train our radio-controlled camera dogs in such a way that we build up expectations in them. So, it's not just about following the directional cues – 'left', 'right' and 'straight ahead' – it's also an exciting game in which we try to convey to them that somewhere out there, far away from where we are, something great will happen. For example, they may find a hidden feeder that suddenly starts dispensing food. Or maybe a ball machine is hanging in a tree or on a wall. It may be that the dog has to work for half an hour with many changes of direction when, all of a sudden, there is a bridge signal and a reward at the handler's position. Or the handler may have changed location, and the dog has to follow a new set of directional cues. In all these scenarios, expectation is the driving force.

Building expectation is an important aspect of animal training. Scent detection dog trainers are heavily dependent on it as novice dogs start out by finding lots of target scent every few minutes, but gradually learn to work longer and harder before they find anything.

Negative expectation can be a serious problem which is difficult to overcome. When scent detection dogs are in training, they have the expectation of finding a target scent every 5 minutes or every 50-100m. Most of the time, this is the case. As trainers, we like to see this happen; we can check the dog's work, we have proof that he is doing his job, and we can reinforce his finds. We can also show bystanders how well our dog is doing. Happy days for both trainer and dog.

But real life – and real-life operations – are different. Often, the dog will find nothing – this is particularly the case with explosives dogs. So, the trainer who has created a high level of expectation in

DISPELLING MYTHS

Dog handlers often tell us that a scent detection dog can only work for 25 minutes at a time before needing a break. But is that true? In the field, gundogs work hours, and explosives detection dogs in the US military search for eight hours or more. After my initial training as a tracking dog trainer, I was convinced that a tracking dog could only follow a track that was one hour old, over a maximum of 1000m. But then I worked with dogs in Scandinavia who could track 10 km!

You can get very creative in scentwork training.

training, and then goes into operations where nothing - or a lot less - is found, is creating major problems. A dog will be quick to distinguish between the training environment and the operational area and, without the high level of expectation, his performance will diminish. He will appear to be searching, but he will miss most of the hides.

A solution might be to hide target scents in typical real-world operational environment, but that is not always easy. Depending on the substance, it can pose a risk to the environment or trigger other security systems, such as electronic explosives detection. There is also the risk of an uninvolved third party removing the search object which, if we are dealing with drugs or explosives, could be a serious security threat.

There is an interesting study by Porritt, Shapiro, Waggoner, Mitchell et al. on the differences in training and work contexts and the performance of trackers, entitled, *Performance decline in search dogs in repetitive tasks and mitigation strategies*, (published in *Applied Animal Behavior Science* 166 (2015) pp. 112-122.) The study shows that a harmless target object can be used for this purpose. In the study group, 21 dogs were trained to find four different targets, three real and one harmless. After this initial training, the dogs were divided into three groups and then sent to their operational environments for further training over a six-week period:

Group 0T (zero target) had to work, i.e., search, but did not find a single target within six weeks.

Group 1T (one target) only found the harmless target and were rewarded for it.

Group 3T (three targets) found the three real targets and were rewarded for doing so.

All three groups also completed two refresher courses within the six-week period outside the operational environment. When they were presented with the four targets in the final test it was found that the success rate in the 0T group had decreased significantly but it was unchanged in both the 1T and the 3Tgroups. So, building an expectation of finding something of interest ensures that performance remains at a high level and, for this purpose, we can make use of fake, harmless targets. *For more information, see novel odour training aid (NOTA), page 173.*

Chapter 11:
THE USE OF REINFORCERS

We have touched on reinforcement as a method of training (see *Chapter Eight*), but its correct use, particularly in relation to schedules of reinforcement, and making use of a different types of reinforcer, is worthy of further study.

To start at the beginning, let's examine the process of reinforcement as a trainer rewards a dog's behaviour:

- The process starts with a bridge signal, such as a clicker, which means the reward is on its way.
- The trainer's body movement informs the dog as what kind of reinforcer is coming. For example, the trainer's hand moves towards the bait bag.
- The trainer opens the bait bag, which makes a familiar sound (associated with food).
- There is a pause while the trainer digs into the bait bag.
- The trainer's hand, holding food, emerges and is presented to the dog.
- The dog approaches the hand and smells the food.
- He opens his mouth, takes it in, tastes, chews and swallows it.
- The treat enters his stomach and the digestion process begins.

So, reinforcement has a beginning and an end; the process is not confined to putting a treat in the dog 's mouth. The food could also come from an automatic feeder, or the dog could be rewarded with toys or a social reinforcer, such as petting, or simply being with his trainer, which can be a powerful reinforcer in its own right.

There are also times when a dog appears to go that extra mile, not for a tangible reward, but because of the relationship he shares with his handler/trainer (see *emotional bank account, page 114*). All of this makes it clear that reinforcement is not simply a matter of giving or withholding. It goes far deeper and has important implications. According to the scientific definition, *reinforcement increases the likelihood that a behaviour will occur.* We can give limitless treats, or play with the dog all day, but it is only an increase in behaviour that tells us that reinforcement is taking place.

Scientists, including B. F. Skinner, have done a lot of work on reinforcement schedules, testing their theories with experiments in controlled environments. Skinner's book, *Behavior of Organisms*, gives some fascinating insights, but it is not an easy read, and can seem a far cry from everyday life where we are working with police/military dogs or with performance dogs. We learnt a lot from these scientific studies when we designed the training plans for our radio directional camera dogs, and for the target training with lasers. However, I prefer to talk about reinforcement at a level where we can reach and motivate the majority of people. Yes, there are no end of complicated reinforcement schedules out there, and yes, I strongly encourage further investigation. But, for now, let's focus on the basic information.

We have already seen that a reinforcer motivates behaviour. It's simple:

if the animal offers more of the desired behaviour, and the behaviour is stronger, the reinforcer is working (see *Chapter Six: The Power of Conditioning*): **Positive reinforcer (R+)** is when there is a positive consequence for the dog and he is given something he likes, such as food and play. **Negative reinforcement (R-)** is when the dog is trying to put a stop to, or avoid, a negative consequence, i.e., something he doesn't like, such as an angry dog owner jerking on the leash.

Both types of reinforcement lead to more and stronger behaviour – dogs learn because their behaviour is modified by its consequences. When the trainer is able to control these reinforcers, by giving or withholding them, they hold the key to behaviour change. So which type of reinforcer should we use?

Imagine we want to teach a dog to jump on a platform. We can use positive reinforcement and feed the dog when he jumps on to the platform. Alternatively, we can pull the dog by the leash towards the platform and release the tension on the leash once he jumps on to the platform. This is an example of negative reinforcement. Either will work, but we would opt for positive reinforcement because it also acts as a deposit in our emotional bank account (see *page 114*). Our relationship with the dog just keeps getting better. But there is a catch. The use of negative reinforcement gets to the goal faster – but only with that particular platform, in that particular environment, and when the platform is in that particular position. As soon as we change the size or colour of the platform, or move it, the negative reinforcement must start all over again.

The great benefit of positive reinforcement, in addition to relationship building, is that the dog is motivated to try new things and show initiative. A platform that looks different, or is located somewhere else, is now his window to a possible reward. He will run and jump up to see if this platform will also lead to positive reinforcement. The dog is looking at the world from a more optimistic angle – and that's exactly what we want.

A major advantage of positive reinforcement is that it encourages dogs to try new things in anticipation of a reward.

To make the best use of reinforcement, we need to work with the two principal categories: primary and secondary reinforcement. To recap:

Primary reinforcers are 'unconditioned' or 'unlearnt' reinforcers that are evolutionarily based, i.e., they have an inherent value determined by basic survival needs – primarily food, water, sleep, and also sex. In addition, there are other primary reinforcers, such as protection from the elements, comfort, community and play (which is a social activity and is also used to develop hunting skills) which are less clearcut. However, when it comes to training, food stands out as the primary reinforcer for the following reasons:

- It keeps a dog in low drive so he is able to learn new things without becoming over-aroused by the reinforcer.
- It is easy to handle.
- It can be consumed quickly allowing the dog to move on to the next exercise.

Play is useful as a primary reinforcer,

but it depends on the task:

- Play often produces high arousal, which may mean the dog is no longer able to process and learn.
- It takes time to introduce a toy, play, and then end the game, which limits the number of repetitions that can be asked for in training.

Rewarding a dog with water would be pretty complicated, and other primary reinforcers, such as comfort or shelter, cannot be put into practice.

Secondary reinforcers are 'conditioned' or 'learnt' reinforcers. They are initially neutral and only develop their power and effect when combined with a primary reinforcer. The sound of a clicker or whistle, a ball, the laser signal or a word of praise mean nothing to the dog at first, but when combined with a primary reinforce, such as food or play, they become significant.

The clicker is a bridge signal that tells the dog a reward is on its way.

Tertiary reinforcers are used as a 'keep going' signal (see *page 156*). It is the feedback that tells the animal that what he is doing is right and that he should keep doing it. The tertiary reinforcer builds motivation and encourages the dog to exhibit the desired behaviour over an extended period of time. In addition, it can be linked to various states of mind (see *mental anchors, page 118*), and it can serve to predict which reinforcer is coming.

REINFORCEMENT SCHEDULES

Our dogs have to learn many different behaviours. To make these both strong and reliable, we do thousands of repetitions and use lots of positive reinforcement. But if we rewarded every behaviour with the same reinforcer, at the same time, from the same distance and position, our dogs would lose interest very quickly. We therefore need to build in an element of surprise which will help to keep them engaged in the process. Of course, we can use different food rewards, such as cheese, chicken or liver, and we can introduce different toys, such as a ball, a tug toy or a kong. Varying the reinforcer has its uses, but the key element in making a behaviour super strong, reliable, predictable, and building longevity, is implementing a reinforcement schedule.

Reinforcement schedules come in many different forms; some are better suited for training in laboratory conditions and others are better for working in the field. Here I will focus on giving an overview of the main reinforcement schedules used in dog sports and when working with utility dogs in the police, military and customs. Firstly, we need to set some rules about when and how to reinforce. For example, do we need to reinforce each time the dog performs the desired behaviour in response to our signal? That depends on our final goal and the level of our training – ultimately it is the rate of reinforcement that conditions the behaviour. It affects:

- The speed at which the behaviour is delivered (fast or slow).
- Duration of behaviour (long or short).
- Strength of behaviour.
- The number of repetitions before the animal stops showing the wanted behaviour. Scent detection dogs can be rewarded for every find because the handler is nearby. Radio directional camera dogs often work at great distances from their handlers and cannot be reinforced for every change of direction.

CONTINUOUS
Learning new behaviours
This involves rewarding every correct response; it is the schedule we use when we want to teach a new exercise. Take 'sit' for example. We lure the dog into a sitting position, and he gets his reinforcer. We advance quickly and the dog learns to expect a reward with every attempt. When rewards are no longer given, the dog stops exhibiting the behaviour, but starts again as soon as rewards come back into the equation.

Obviously, this has its limitations. We don't want the 'sit' behaviour to be dependent on the dog's expectation that he will be rewarded for every correct response, or that he will only sit when the reinforcer is present.

When I am working with a scent detection dog, and training is progressing, I continue to reward every time he finds something, but I vary the type of reinforcer to keep things interesting - sometimes food, sometimes play, etc. Bear mind that verbal communication, saying *"yes"* or *"well done"*, is a reinforcer connected to social activity (see *page 113*).

INTERMITTENT
Showing behaviour even when the reinforcer is not present
In this situation, we no longer reward the 'sit' every time, but only *sometimes*. When the dog sits in response to our signal, he sometimes, but not always, gets a treat, so he will sit even when the reinforcer is not present. The advantage is that, as soon as the new behaviour is understood, we can build up a higher level of expectation than with continuous rewards. It is more exciting for the dog: *"Do I get a reward or not, and if so, which one?"* When we train a new behaviour, it is important to get away from continuous reinforcement as quickly as possible. There are different schedules of intermittent reinforcement for different purposes:

Fixed rate
Creating strong behaviour chains
In this schedule, we always reinforce after the same number of correct responses. For example, if we work with the fixed rate of five, we reward every fifth 'sit'. This pattern is easy to set and follow, and easy for the dog to remember. It's useful for building a continuous behaviour, such as weaving, or even an entire agility course. For example, we can set a fixed rate of two and only reward every other course run. This reinforcement rate is also used in the human workplace, for example, workers get a break after they have packed 50 products on an assembly line.

Variable rate
Keeping going with minimal reinforcement
This involves planning an unpredictable reinforcement rate before we start training so the reinforcer appears randomly, as

far as the animal is concerned, but there is still the guarantee of a reward. I like to use the Google Play app *Pick a number!* to help, depending on my training plan. For example, I might consider doing 15 trials but only reward four of them. In this instance, I would enter one as the minimum and 15 as the maximum in the app and click four times on generate number. These trials – for example, 6,8, 9 and 15 – will be rewarded once they are sorted into ascending order. The same principle applies to slot machines; wins are unpredictable but what is certain is that there will be at least one or more wins per gaming session.

The advantage of this schedule is the animal is motivated to keep working because he never knows when the next reinforcer is coming, so he really tries his best! It also ideal for teaching dogs to keep working over a long period of time, especially if there is a reason why the reward is delayed. For example, a variable gain rate of 1000 was used to train the first monkeys in space. They had to show 1000 responses before they got a reinforcer. Of course, this would never be a starting point, but if the animal becomes accustomed to a progressive schedule of variable reinforcement it can be built up over time.

FIXED INTERVAL

Not recommended for dog training

This involves fixing an exact period of time before the animal is rewarded. For example, a fixed reinforcement interval of one would mean that the dog would always be reinforced exactly one minute after he responded to the signal to sit. However, dogs have a very good sense of time, which can be counter productive in training. For example, if a dog is routinely fed at 8 am and 4 pm, he knows exactly what will happen at these specific times. He has no curiosity, and no need to think, because his mealtimes – his reinforcers – are entirely predictable. This schedule in used in laboratory situations, in a controlled environment, but it is definitely not recommended for dog training.

VARIABLE INTERVAL

Ideal for delayed reinforcement

To explain this schedule, let's consider how we transport dogs; whether it be it by car, boat or helicopter, we use crates. This is not only for the safety of the driver/captain/pilot, but also because it provides a place where the dog feels safe. No matter which means of transport we choose, the crate is a familiar, known quantity. This sense of security, associated with the crate, is trained with a variable reinforcement interval. We gradually extend the time the dog is in the crate, breaking up the time he is confined with a fun game. We keep building this up until the dog is happy to stay in the crate for an extended period of time.

We also use variable intervals in scent detection work. Initially, a young dog has a find every few minutes. But then we plan the hiding places so he has to search for`;

- 2 minutes before a find
- then 6 minutes
- then 1 minute
- then 12 minutes
- then 23 minutes
- then 5 minutes ...

The time, i.e., the interval between the finds is variable, and that is exactly what keeps the dog searching. However, as with all reinforcement plans, it is important to control what is happening by collecting data and evaluating it.

FIXED DURATION

Training to stay focused on something or someone

One of the biggest challenges for attack dogs is to stay in a covert position and remain silent and motionless. If we use a fixed duration schedule we can train them to stay in position for a fixed period of time. If the dog meets the criteria, say three minutes, he will be sent to attack. If he doesn't meet the criteria and does something else after two minutes, such as getting up, barking, whining, etc., he will not be sent out so he misses out on all reinforcement. We also use this schedule for our radio directional camera dogs. One of their jobs is to look into a window for a set period of time (so the camera can film what is happening inside), or to look at an object for a fixed time.

Our camera dogs learn to look into a window for a variable duration – and the reinforcer is only available after this time lapse.

For example, we set a fixed duration of one minute (FD1). If the dog watches/films for a minute and does nothing else, we reinforce. If he misses the minute and looks the other way for a brief moment, there is no reinforcer and we start again. Over time, we increase the duration. As with any training, it is extremely important to have clear criteria for both the dog and the trainer, otherwise it just won't work. If the dog fails to meet the criteria, resulting in a reset, both dog and trainer need to understand what exactly caused the reset and why there was no reinforcement.

VARIABLE DURATION

The schedule for radio directional camera dogs

This schedule is key to training radio directional camera dogs for operations. The dog has to work for an extended period before he gets a reinforcer, but because the time intervals between reinforcers are always open to change, he cannot predict when the reward will come. The difference between *variable duration* and *variable interval* is the type of duration that is measured:

Variable interval measures the variable time-lag from reinforcement to reinforcement.

Variable duration measures the time the dog is working, which is variable.

For example, a dog may have to walk in the open for two minutes, move through a building for five minutes, and then wait ten minutes before he gets his reinforcer. The same principle is used to train scent detection dogs: search a small room, then search a large room, search a smaller room, search two large rooms, and so on.

THE THREE WS

Help for young trainers

I like to encourage young trainers to keep their training both exciting and interesting. The dog only needs to know one thing for sure – that a reinforcer will come at some point – and he must find out how to win it! This is the magic of training. Once a dog has mastered the basics of a new behaviour, we make sure he *never* knows:

- **WHEN** a bridge signal and/or reinforcer is given. This means that while working, the dog is always surprised when a bridge signal or reinforcer is given.
- **WHAT** the reinforcer will be. If a trainer varies the reinforcers, there is always an element of surprise as the dog cannot predict whether it will be food (what type of food?) or play (what toy/what game?).
- **WHERE** the reinforcement will be given. Another element to keep the dog guessing: will it be from a ball machine, or an automatic feeder? From a third party, or from the trainer?

By playing with the three Ws, training is never routine and predictable. It is full of surprises, which keeps the dogs fully engaged and interested.

A FLEXIBLE APPROACH

American behaviourist, David Premack, who worked extensively on the use of reinforcers in training states that anything in the environment can be used as a reward. So reinforcers are all around us. A reinforcer could be rolling in the fragrant grass, swimming in the pond, digging a hole in the sand, doing zoomies, or jumping into a pile of leaves in the forest. Sometimes we trainers are simply not smart enough when it comes to choosing reinforcers. We need to be creative and keep an eye open for possibilities. And remember, it's not just about what reinforcer we use, it's about how we present it. We need to mix things up: throw it, give it, let the dog look for it, play with it, hand it over calmly or hand it over excitedly. There are so many different options – and they make a world of difference to the dog.

A misunderstanding has arisen among some trainers attending the chicken workshops, where trainers hone their reinforcement skills (see *page 74*). One of the first things students learn is to stand as still as possible when reinforcing (bridge signal + food reward), to use the same movement sequence, and to give no other body language or vocal cues that the chicken might mistake as part of the process. It is good practice to be super aware of unintentional cues (particularly in the case of chickens) but this does not mean that when we move to real-life training scenarios we should always reinforce the dog in a robotic and emotionless way. How joyless for the dog! Instead, it is important to adapt to each particular situation. For example, if I am training a new task that requires a lot of concentration and precision from the dog, such as flicking a very small switch, I give reinforcers very calmly and very quietly to avoid putting the dog in an unwanted state of excitement. Too much movement, emotion and energy would disturb his concentration. In this instance, I might even use an automatic feeder. However, once the dog has mastered his new task, I immediately go back to rewarding him with a voice, movement, a happy face and lots of positive energy!

There is another reason why it is important to vary the type and intensity of the reinforcements we offer. In our covert operations we often find ourselves in situations in which we must behave inconspicuously. Often, after a successful job, we cannot praise the dog out loud or play ball with him; we have to rely on a soft word and a gentle petting, even when his performance has been outstanding. If this kept happening in operations – never in training – we would, over time, end up with a dog whose work was worse in operations than in training, because of the lack of more obvious reinforcement. To counteract this, the intensity of reinforcers should also be varied during training.

Reinforcement through play is not always possible in operational situations. Therefore, we need to vary the type and intensity of reinforcement during training.

The type – and intensity – of reinforcements should be changed and adapted depending on the situation. The secret is to plan ahead!

DIFFERENT TYPES OF REINFORCER

Reinforcers can be anything, so what should we consider in addition to food and play?

SOCIAL REINFORCERS

Social reinforcers can be very powerful, although we are often unaware that we are giving them. A typical example is when the dog has made a mistake and the handler is waiting for feedback or preparing for a reset. The handler, without thinking, pets the dog, giving a social reinforcer for something we *don't* want to reinforce. And for some dogs, these social reinforcers are even more important than food.

We are all familiar with the scenario of the dog who jumps up at people and receives constant social reinforcement. Even if their guardian yells or tries to push the dog away, attempting to stop the behaviour, the dog will still jump up. A touch, eye contact, words, play, or attention are all social reinforcers, and they are very powerful. In my work, we avoid such unwanted social reinforcements by making use of automatic feeders or ball throwing machines. If the dog makes a mistake or, rather, doesn't meet the criteria, the machine will not deliver. This means there will be no reinforcement and, importantly, no interaction with the handler, even if it is as minimal as putting the food, or toy, back in a pocket. For some dogs, the trainer simply stepping into the training environment is reinforcing.

EMOTIONAL BANK ACCOUNT

In my previous job, there was a manager who was very good at asking for, and getting, extra work from his employees. Even on Friday afternoons, when everyone was getting ready for the weekend, he always found someone to finish an important task. Or if there was a crisis, he would find someone to work a night shift and solve the problem by the next business day. And if nobody on the team wanted to work overtime, he persuaded other staff members to step in. He told me that he could call in favours because of the relationship he had developed with his staff. He had set up an emotional bank account, and he could ask for, and expect, withdrawals from it

Although much disputed by behaviourists, I believe that an emotional bank account exists in human-canine relationships. These are the moments when a dog offers a very special behaviour, a very special feeling or a very special connection – things that we cannot train. I am aware that these reports are anecdotal, i.e., unscientific, but I have experienced these moments on many occasions.

I like the idea of filling a dog's emotional bank account by going for long walks in the woods or jogging together and letting him be with me as much as possible. This is precisely why I use so much positive reinforcement in my training and why I am passionate about advocating its use for others. Positive reinforcement allows trainer, handler and guardian to build a strong and reliable bond with a dog – and he will give back so much more in return.

JACKPOT REWARDS

A jackpot means stepping up the reward so it is better than normal - a bumper food reward, or a longer, more intense game of tug. Beware, jackpots should not be given often, and certainly not for simple behaviours. They should be given when the dog has done something outstanding: a lightning-fast recall, completing a tricky handling move, or winning a competition!

In recent research, Kennon A. Latall (*Behavior Analysis in Practice 13*, 1000-2008, 2020) describes jackpots as large, infrequent reinforcers, and concludes that they lack clarity of definition and procedure. I would ask the following question: *Are jackpots given for exceptional performance and do the animals understand that they are special?*

Could it be that a jackpot reward has more to do with the trainer's state of mind - a way of conveying pride and satisfaction in what the animal has just achieved? I believe that this area of training needs more research, but I would caution trainers and handlers to limit the use of jackpot rewards to exceptional achievements.

ACTIVITY REINFORCERS

Following David Premack's premise that reinforcers are everywhere, we need to examine the importance of the activity itself, and whether it can be used as a reinforcer. One of the questions we are often asked in scent detection dog training is should we reinforce the dog after he has searched a designated area, such as a room, before proceeding further.

CHART TOPPERS

When choosing reinforcers, it's important to know what a dog really wants and values because, as already highlighted, he is working on the principle of *'what's in it for me?'* (see *page 49*). We, therefore, need to ensure the dog is motivated to work for us. One of the key elements in maintaining a dog's interest is variation. I have come across trainers who, when assigned with a new dog, simply continue to use the reinforcers they used with their previous dog. In this instance, the trainer is deciding what is reinforcing for the dog. Yes, most dogs like a tennis ball, a kong, or some food, but is that reinforcer producing maximum focus? We need to find the best, and most valuable reinforcer for the dog we are working with, include some variation, and ensure that the reinforcers are appropriate for the situation and level of arousal (see *page 141*). We should also give the dog a reward of lower value for an easy task and a high value reward for an exercise that is more challenging.

Suiting the reinforcer to the situation is something we humans can relate to. Imagine a 12-year-old boy is training with his buddies for an important football match. The boy rushes forward and scores a goal – and his mother, who has been watching, runs on to the field, kisses him, fondles his hair and gives him a lollipop. Will he like that? Certainly not! Even if he usually likes lollipops... Dogs are no different!

When I am working with handlers, I recommend making a list of the top five reinforcers for each individual dog. We determine this in a very simple way. We ask someone to restrain the dog and then place five reinforcers in front of him at a distance of approximately 2m. These are assigned numbers, from left to right, for example:

- Kong
- Tennis ball
- Dry meat
- Kibble
- Fresh meat

We release the dog and observe what he chooses. If he chooses number 5, the fresh meat, we repeat the whole thing: the dog goes back to the starting position and we put a fresh piece of meat at station 5. If he chooses 5 again, we know that this is number one on our hit list – it is the best reinforcer for this dog. We are now ready for the next round, this time leaving station 5 without a reinforcer so the has only four reinforcers to choose from. Which will he choose this time? We repeat the exercise and this gives us number two on the list. This process is continued until we have a list of the top five (or more, if required) – reinforcers for that particular dog.

Scent detection dogs find the activity of searching highly rewarding; the same is true of tracking dogs who sometimes overshoot or ignore the object on the track out of sheer eagerness to search. They want to keep going because they enjoy searching so much! Agility dogs also get massive reinforcement from having access to the equipment and the opportunity to keep on running.

We can make good use of activity reinforcers to give value to more mundane exercises such as a down-stay. This exercise should be built up gradually and rewarded with something the dog really loves – a game of tug, access to a specific piece of agility equipment, such as a tunnel, a food scatter, or simply the opportunity to run free. This will build a positive association with the down-stay exercise, and the dog will become more willing to co-operate.

In the human world, we often apply the Premack principle, without being aware of it. For example, one of the kids in our neighbourhood loves washing his dad's car. I asked him how much money he got for it, because I thought about asking him if would like to wash my car, too. He happily told me that his 'payment' was being allowed 30 minutes extra play time on his Xbox!

Ygor gets a great reinforcer after a perfect 'sit'.

NO REWARD MARKER

Now we need to look at how to deal with situations where no reinforcement has been earnt, but the dog is expecting it. This can create real conflict for the dog and can test the trust that exists between dog and handler/guardian. There are a variety of circumstances that can lead to a situation of no-reward. For example:

- During scent detection training, the dog alerts on a specific object. He seems convinced that it's the right one, focuses on it and freezes. Just at the moment when he is expecting to be reinforced, his handler receives information that it is a false alert and that the object does not contain or carry the target scent. What should happen next?
- What if he finds the target scent but instead of a passive response, he starts scratching or barking at the object? Should he still be reinforced?
- Or what about the dog who responds to a series of cues and signals in obedience training, but makes a couple of mistakes before the end of the sequence? He is expecting a reward, but the handler has decided to withhold it because, as we know, we always get what we reinforce.

In all these examples, the dog 'believes' that he has done the right thing. As handlers, we know why we are not giving reinforcement – but is it clear to the dog? Why is he not getting reinforcement – is it because he has made a mistake, or are we delaying the reward to build duration? Obviously, this is both confusing and frustrating for the dog, and underlines the importance of clear communication.

Animal training communication generally focuses on letting the animal know he is doing something right, and for this we use a keep-going signal, a bridge signal, food, play, body language, and so on. But we also need to to tell him when he is doing something wrong, in order to enhance his understanding and to speed up training. When I ask trainers what they do when their dog meets the criteria, most know what to do: bridge and reward. But when I ask what happens when their dog makes a mistake, the response is muddled. While most have a vague idea of a desired outcome, they are unclear about what to say, how to respond, and what steps to take.

American researcher, writer and speaker Brené Brown provides a good insight on communication – *"Clear is friendly, unclear is unfriendly"* – and this applies equally to dogs.

Consider the example of learning to drive. It helps to get a lot of feedback from the instructor, not only when things are going well, but also when things go wrong. Imagine you are trying to do your best, but the driving instructor doesn't say anything. Then, at some point along the route, he suddenly slams on the brakes and tells you to go back to the starting point and do the route again. His silence is unsettling, and he has given no information as to what you may have done wrong. Clear communication – including what we are doing wrong – helps to define the right behaviour. This is essential for the learning process.

Not every mistake is the same. A scent detection dog working a line-up may give a false positive alert (on a distraction odour) or a false negative (target scent missed). An explosives detection dog can alert on a bag in the airport where nothing is found. A dog in obedience training may break his 'down' position and walk up to his handler. An agility dog may get drawn into taking a tunnel, which is an off-course obstacle. Silence is not the best form of communication as it provides no feedback. We are very ready to use a bridging signal to mark good behaviour, so why not use a signal to tell our dog that *no* reward is to be expected? This is known as a 'no reward marker' (NRM) or a 'negative marker'. We don't want to attach punishment or conditioning to this signal, it just lets the dog know that, from this point on, he's wasting his energy and he will not get a reward.

A no reward marker is most effective when it is a verbal cue; we always have it with us and, unlike a specific item, we cannot forget it. When using a verbal cue, we can also vary volume, pitch and length. We say a quiet, friendly, *"no, no"* at the exact point where the dog fails to meet the criteria. This provides clear communication, and the dog gets further help in the form of a keep going signal (see *page 156*), or a reset back to the start. Here we can make use of a dog's ability to respond to our words and our intonations. This gives us the ability to use a louder or sharper *"no!"* to stop a potentially dangerous behaviour – for example, when he's about to run towards a busy road, step on broken glass, or consume something that is toxic. In certain circumstances, a negative marker can be used in training, putting the dog under a degree of mental pressure to stop what he is doing. However, this needs to be built into a training programme and its effectiveness depends on the relationship between dog and trainer/handler. We should always use these procedures with caution, setting the dog up for success, proceeding in small steps, and analysing the resulting data.

MENTAL ANCHORS

When we talk about reinforcers, reinforcement schedules, and non-reinforcement markers, our goal is to get, or rather capture, the desired behaviour by giving or withholding rewards. The ultimate goal is to train our dogs to focus on work, ignore distractions, and give us the behaviours we want. All of this can be difficult in an environment full of distractions. In our human world, dogs are constantly surrounded by sounds, smells, people, traffic, other animals and much more. Mental anchors can help them stay focused. They serve as a point of reference in a hectic, stressful environment. It is like a friend, who we respect and trust, putting an arm around our shoulders at times of stress. We no longer feel alone; we find a moment of peace and we can concentrate on the next step.

Todd demonstrates the mental anchor.

Camera dog, Ops, prepares for action.

We condition appropriate mental anchors for our dogs because they cannot learn if they are too excited, or in too low a mood. We need to get our dogs in the right frame of mind before we ask them to work. It is more than a starting ritual because it depends on the individual dog's sensitivity to the training environment.

Our operational dogs work in very intense situations, often within the vicinity of gunshots, explosions or people who are shouting or screaming. It is, therefore, especially important to give them an anchor point before they are sent out to do their job. It's like us closing our eyes for a brief moment and taking a deep breath. There are several ways to condition a mental anchor. My preference is to offer an open palm to the dog. On cue, the dog rests his chin on my hand, and anchors there for a few seconds before starting work. Another option might be using a platform or a crate (with the door open).

Chapter 12:
THE LEARNING PROCESS

There is no strict definition of what constitutes learning. We can agree that it involves a change in behaviour and so, if there is no change in behaviour, the individual involved has not learnt anything. However, we need to be careful here; the change in behaviour must be repeated after the learning phase. If the individual shows the changed behaviour on a single occasion, we cannot be sure that anything has been learnt. Along with learning, teaching can also be difficult to define because there is no consensus on what constitutes good teaching practice. There are many animal trainers out there, but are they capable of coaching other trainers? Are they capable of getting the best from the dogs that come to them for training?

A teacher's role is to transmit information. Good teachers make us curious to find out more – to find out what is possible. A good teacher presents us with challenges, leading the way and encouraging us to step out of our comfort zone. Such a teacher can make a tremendous difference to their students. Teaching and learning is not just about putting information in students' minds. More importantly, the student needs to learn what that information is needed, thereby facilitating success in the operation, the competition, in the given situation or task.

In general, we can distinguish four phases in learning. Let's take the 'sit' behaviour as an example, using my preferred training method, luring with food. When we first start teaching a dog to sit, he has no idea what to do. To achieve the task, he has to work through four phases of learning:

Learning a new behaviour

In this phase, the dog is introduced to a new behaviour. He has to figure out what we are asking for, and how to respond.

To start the learning process, I pick up a tasty piece of food, hold it over the dog's head, and lure him into the sitting position. As soon as his rear hits the ground, I open my hand. The dog gets the food and starts to understand how to 'open' my hand to get the food.

Mastering the behaviour

The dog understands what he must do. He reacts faster, more intensively and stays in position for longer. It's like an aha-moment – he has mastered the behaviour and can perform it fluently.

Luring with food is no longer needed; the dog understands what to do. He starts to anticipate the behaviour, which is the moment I can put the behaviour on a cue.

Generalisation of behaviour

The behaviour is transferred to different locations.

Now I ask for the 'sit' behaviour in a variety of different situations. Instead of practising in the living room, we go to the garden. We practise the 'sit' on different surfaces, and then on a platform, or a box. In short, I do everything to ensure that the dog learns to sit under all conceivable circumstances and with all possible distractions.

Maintaining the behaviour

Ensuring the quality of the behaviour is maintained

Finally, we look at the overall performance and work out how often we need to train the 'sit' to ensure the dog continues to respond quickly, reliably, and for the required duration.

SAFE LEARNING ENVIRONMENT

Before training gets underway, it is important to create a safe learning environment. The atmosphere in the classroom and outside on the training ground has a major impact on students, dogs and handlers alike. I have seen many training settings around the world, and I have come to realise the importance of creating a positive environment. If the learning environment is burdened with ego, shame, or guilt, it creates tension that is easily perceived by all participants.

We need to ensure that everyone is feeling good, safe and ready to learn. We should consider the fact that our brains associate places with specific activities. For example, if we get into the habit of working from the sofa in the living room, we start to associate this room with working and it becomes difficult to relax at times when we are not working. The same also applies to animals. When we create a learning environment, it should be filled with positive emotions and not be associated with other activities, such as routine care or grooming.

There are also some practical considerations which will assist learning. Dogs should be exercised and have the opportunity to relieve themselves before training. They should also have access to fresh water within the training environment. Trainers/handlers should prepare a training plan in advance and set

A safe learning environment refers not just to the place itself, but also to the atmosphere which should be without pressure and tensions.

up equipment before the start of a session. Mobile phones should be switched off, and it is important to ensure there will be no interruptions when training is underway. *For more information see Chapter 20.*

MENTAL PREPARATION

We have all experienced the influence our mood, or state of mind, has on what we do, and training is no different. After an awkward phone call, a nasty argument with the boss, or a fight with a partner, training rarely goes according to plan. To be successful, we need to optimise our training time to ensure we can achieve the state of mind that ensures we can make progress. To do this, we not only need to prepare the environment, targets, reinforcers, and have a training plan, we also need to make some mental preparation. We need to question whether we are in the right state of mind to start training. Are we open, happy, balanced, focused and at ease?

Our dogs also need to be in the right state of mind to be successful in training. We have to make sure that the dog is not only in good physical shape, he also needs to be in a good place mentally. I believe this starts as soon as we wake up in the morning, giving our dogs the attention they need. I always go for a good walk in the morning, giving my dogs the opportunity to investigate what has been going on in the neighbourhood by sniffing each other's scent tags. They enjoy exercise and being outside, but also attention and social contact. For me, it is also part of everyday life to ask for obedience before feeding and, when required, to solve tasks together. For example, I teach my dogs to settle when I am busy, and I introduce them to all possible means of transport, car, train, bus, etc. At an early stage, I expose my dogs to distractions that they may encounter later in training, or in action. This could be learning to focus in the presence or rabbits or squirrels, ignoring discarded food on the

A TIME TO FOCUS

Without attention there is no learning process. Asking a dog to focus so he can attend to the cues and instructions he is being given seems straightforward. But there are so many distractions – things the animal needs to process – which means that paying attention is a big challenge. It's not just about signals given by the trainer; the environment provides many signals in the form of sounds, smells, temperature, wind, touch, rain, soil conditions, contact, colours, other animals and much more.

In addition to cues from the trainer and the environment, the animal's motivation also plays a role and, as we all know, this can vary and change very quickly. Aggressive dogs are on the lookout for a fight, anxious dogs are busy trying to avoid trouble, hungry dogs are preoccupied by food, sociable dogs are attracted by other dogs and by people. These, and other circumstances, change motivation and attention. For this reason, we need to set up a training environment in a way that allows the dog to give maximum attention to the task in hand, especially early on when he is learning a new behaviour and is struggling to process all the distractions around him.

street, working in heavy traffic, on different surfaces, or learning to negotiate obstacles. We learn from each other, discovering both strengths and weaknesses and, as an added bonus, my dogs will be learning to read my body language.

We don't have to prepare our dogs for the learning process because nature has equipped them so that they are ready to learn around the clock. If an animal was not willing and able to learn, it would not survive long. But we can prepare both our dogs and ourselves for a training session. We need to consider the following:

- What are the dog's favourite reinforcers? Are they to hand?
- Are we able to change the tempo of learning, switching from high-drive learning to low-drive learning and back?
- Do we understand the dog's body language and does he understand ours?
- Is the bridge signal properly conditioned?
- Is the training environment properly prepared?
- Are training protocols prepared and within reach?
- What is the focus of the training session?
- What about gathering data?
- What about a plan B, if equipment malfunctions?

The aim is to achieve maximum results, making our training truly valuable.

SOCIAL LEARNING

This type of learning means that the animal is motivated to imitate a behaviour shown by another living being. Edward Lee Thorndike, a pioneer in operant conditioning, conducted a number of studies but was unable to prove that dogs learn through observation.

Ready for training.

However, in 1996 psychologist, Jacques Vauclair, provided scientific evidence that observational learning is possible in many different species, including apes, birds, dolphins, and dogs. Today, in dog sports around the world, we see a lot of handlers working according to the *Do as I do* principle pioneered by Claudia Fugazza. In her book, *Do as I Do: Using Social Learning to Train Dogs*, she explains how fun and easy it is to train dogs in this way.

Social learning is nothing new; it is one of the basic elements of survival. Puppies mimic the behaviours of their parents and their siblings in hunting games as they explore the world around them. We use mimicry when we are training a new skill, such as touching a target. If we focus on the target and touch it, the dog will copy our behaviour. If we then shape his attempts by marking approximations of the desired behaviour with bridging signals, we very quickly reach our goal.

However, Thorndike, and other

scientists, have a different view of social learning. Imagine training 'touch the target' using trial and error. When the dog has learnt the behaviour, he is put in a room which is divided by a panel of glass. On the other side of the panel, a dog is introduced who has never seen targeting behaviour. The untrained dog watches as the trained dog touches the target. However, unless the untrained dog receives help from a human trainer, he is unlikely to master the behaviour and reproduce it exactly, simply by watching the other dog.

This is where the paradigms of learning by observation between scientists and dog trainers collide. Of course, dogs look at us trainers, but they also look at other dogs. They copy parts of the behaviour they see, but they will never reach the outstanding level of mimicry that we see in marine mammals. I have seen the most impressive examples of social learning among dolphins, who are able to reproduce the exact behaviour of other dolphins just by observing and communicating with each other – without the help of a trainer. At dolphinariums in the Netherlands and Hawaii, I observed the speed at which they learnt from each other. If we asked a dolphin to show us a few random behaviours and reinforced them, another dolphin would be able to reproduce the identical repertoire, without any cues, within a matter of seconds. The power of learning by observation was further highlighted when an orangutan in Borneo was photographed using a spear to catch fish. There were no known reports of human teaching orangutans to perform this task – it could only have come about by observing and then copying human behaviour.

However, social learning is not always positive; animals can also learn to feel stress or fear by watching others. If one dog is afraid of something, another may copy his behaviour. Equally, if a visiting dog raids the bin, or digs holes in the garden, don't be surprised if the resident dog picks up some new ideas!

ERRORLESS LEARNING

We always try to set up our animals for success, meaning we design our training and our training environment in such a way that the animal will be successful. Help, and the use of cues, are gradually phased out so that mistakes are avoided. In this way we achieve a situation of so-called 'errorless' learning. But beware, the learning process also consists of trial and error, which is why we need to pay a little more attention to this topic.

Psychologist Herbert Terrace conducted an interesting study on pigeons in 1963 which involved teaching them to distinguish between red and green. First, the pigeons were trained in a conventional way: they learnt to peck a red dot and received a food reward for it. When they were successful in demonstrating this behaviour, they were presented with a green dot. Initially, they pecked at the green dot, but they did not get any food for it. Gradually they pecked the green dot less often because no reward followed. This is the conventional way of training a discrimination task. In the study, the pigeons pecked the green dot frequently – 3000 times in 28 training sessions – before the behaviour disappeared.

In the second part of the experiment, the pigeons (this time, different birds) were given help: the visual stimuli (red and green dots) varied in colour intensity and the length of time they were shown. For example, the red dot appeared bright red for 5 seconds, while, initially, the green dot appeared for only 0.5 seconds and was very dark. Only later was the exposure of the green dot increased to 5 seconds, and the colour intensity strengthened so it was bright green. With this extra help, there were only 25 incorrect responses to the green dot in 28 training sessions. From these findings, Terrace put forward the premise that mistakes are not necessary for learning. Following further research, conducted in 1972, he showed that errorless learning also eliminates behaviours borne of frustration, which can occur with conventional discrimination task training. As a result, training never becomes aversive and performance improves over the long term.

However, for errorless learning to be effective, we need to be able to manipulate the environment and have full control over all forms of stimulus. In the world we are operating in, these controls cannot be put in place, unless we are working in a laboratory situation. So sooner or later, our dogs will make mistakes because situations change, and the impact of unforeseen stimuli is unpredictable. But, if through a process of trial and error, a dog learns to cope with making mistakes, he will develop the perseverance to keep going in order to find a solution that will produce the reinforcer. It's like riding a bicycle: the novice rider needs to make mistakes, fall down and try again to find their balance.

LEARNING BY CONSEQUENCES

If we understand that learning takes place through positive and negative consequences, we also need to appreciate that animal training cannot be designed so that there are *only* positive consequences, or *only* negative consequences. A consequence is a result or an effect that can be pleasant or unpleasant. However, both can be related to each other. For example, if we go to the dentist with toothache and the dentist extracts the tooth, it is uncomfortable at first. But a few hours later, our pain is less than it was before we went to the dentist. Now the unpleasant consequence changes into a pleasant one! Consequences are not black and white; it depends on how the individual experiences them:

- What is the previous experience and current condition of the recipient?
- What is the context?

All of this plays an important role in the learning process.

In animal training we refer to the 'black box'. A stimulus (or signal) goes in, but the behaviour that comes out is not always as we anticipate because we don't know exactly what is going on in the black box, i.e., the recipient's brain. Looking at it from a human perspective, when we go to a supermarket, our buying decisions are influenced by advertising, product placement and price. In addition to all these external influences, our mood will also influence our behaviour – if we are hungry, we will buy more groceries.

Intelligence – the ability to acquire knowledge – also plays a part. Just imagine if we humans had to experience everything on a personal level to find out what leads to pleasant or unpleasant consequences.

It would take us a lifetime to learn how to survive, trying to discover what food is good for us, what makes us sick, how to find a bed for the night, and how to make fires to keep warm - everything that allows us to avoid pain and discomfort, and everything that allows us to experience pleasure. As intelligent people we can learn all this from others and, in turn, pass knowledge on to others.

The use of this type of intelligence is also evident in the animal world. Primates have learnt how to make tools to obtain food by watching each other, and then practising the necessary skills. Magpies are known to observe other birds secreting food, and then return to raid their hiding places. There is also the example of blue tits who learnt how to peck at the foil tops of milk bottles that were, traditionally, left on doorsteps. The birds copied each other and, in no time, there was a whole 'culture' of blue tits in the area who specialised in pecking at milk bottles. For our scent detection dogs, it means learning to remember, recognise and detect odours of interest.

The power of learning by consequences is clear to see.

When it comes to reflecting on the consequences of their own behaviour, some studies have compared dogs to the intelligence level we would expect in three to four-year-old children. Children of this age learn that if they touch a hot radiator, they will experience pain. The next time they approach it and feel its warmth, they will remember the unpleasant consequences and will not touch it. This can also be generalised to other contexts. For example, a child will keep away from a hot radiator in another house, even if it looks slightly different. Now imagine a parent shouting *"no!"* to the child just before they touch the radiator. Because the child does not yet know what *"no!"* means, they touch the radiator and burn their fingers. They remember this painful experience, and not only learn that a hot radiator means trouble, but also that *"no!"* means trouble.

Of course, this 'consequence conditioning' procedure works just as well for pleasurable experiences. If a dog responds to *"here!"* and gets a tasty piece of sausage, his response will be faster next time. Stealing a piece of meat from the kitchen table is pleasurable for a dog and he ignores his guardian's reprimand, *"no!"*. However, if the reprimand and the theft are followed by unpleasant consequences, the dog will learn.

WORKING WITH POSITIVE AND NEGATIVE CONSEQUENCES

I am always troubled by the divisions between purely positive trainers, balanced trainers and those who only work with punishment; arguments and acrimony resulting from this do not help us learn from each other or understand more about animal training. It only serves to polarise feelings and split the animal training community into two camps. Advocates of both types of training go to the barricades to prove that they are right and the others are wrong. This helps neither the cause, nor the animals that are being trained.

It is often the 'converts' – those who previously practised punitive training methods and have changed track – who are the most fanatical in attacking their former colleagues. It is like the ex-smoker who is at the forefront of condemning those who still smoke. Some of the most influential people in the canine community follow the same tactics. They have built their name and company using punishment-based methods to correct all sorts of problem behaviours. Then, when faced with heated debates on social media, they no longer talk publicly about 'problem behaviour', instead they preach the benefits of purely positive training.

Problematic and undesirable behaviour is simply ignored. If a dog shows aggressive behaviour towards another dog when out on a walk, the solution is to take a different route. If a dog fails to respond to the recall because he prefers to chase rabbits, he is kept on a leash. If a dog steals food from the table, he is put in another room. Of course, this 'solves' problems, but it also creates problem dogs. In time, many of these dogs are excluded from their families, confined to a crate, and only exercised for short periods of time. Or worst-case scenario, they end up in rescue and may even be euthanised. Good training, and setting appropriate boundaries, would not only prevent these disastrous consequences, but it would also solve the behavioural problems and give the dog a better life. Setting limits does not automatically mean applying aversive techniques, it's about rules and compliance (see *page 127)*

We also need to give some careful thought to what we mean by 'unwanted' behaviour. Generally, we see this entirely from *our* perspective. It is all about how we perceive behaviour, and how we combine it with our knowledge, our experience, our thinking, or our mood in a given situation. The definition of what is 'unwanted' is, therefore, purely subjective.

I remember a young Labrador driving everyone in his family insane by carrying everything he could find into his basket: shoes, the TV remote control, the car keys and so on. When he eventually started to defend his 'conquests' and also broke an iPhone, the family had enough and gave him to a rescue centre. His behaviour was 'unwanted'. A few weeks later, the same dog was adopted by a police officer who was looking for a young dog to train for drug detection. After only three months, the Labrador was the best drug-detection dog in the unit. So we need to be careful before labelling a behaviour as 'unwanted' – a lot depends on the circumstances, the triggers, and other factors.

But there will also be times when we clearly identify unwanted behaviour. We can choose to ignore it, but if the behaviour becomes dangerous or creates dangerous situations for the dog, for people or other animals, this is no longer possible. Consider a dog who growls at visitors coming to the house. He quickly learns that growling pays off because people get scared and give him what he wants, i.e., to be left alone. Such dogs also learn that it's okay to fight with other dogs out on a walk, and ignoring their guardian is acceptable. They do not react to *"here!"* or to other basic commands and are potentially a source of danger on the road, endangering their own lives, and the lives of others. In these situations, setting limits is a necessity. These limits apply equally to us humans; if we ignore a red light, we get a ticket or die in an accident. Boundaries have a function. A well-trained dog is a joy for the whole family. He responds to basic commands and understands that life has limits.

However, the subject of imposing boundaries triggers heated discussion. For some people it's okay if their dog barks a lot, for others it's a problem. Most people don't want their dog to rush wildly in the woods to chase squirrels, but some don't mind and see it as extra exercise for the dog. There is no 'one size fits all' definition of a well-trained dog because we all have different relationships with our dogs and different opinions. But we can all agree that a well-trained dog should have a basic level of obedience, he should not start fights, steal food from the kitchen or be aggressive towards humans. Such a dog will be a valued member of the family and is not in danger of being excluded. He will be fit for everyday life. In return, his basic needs should be met in terms of food, shelter and exercise. In addition, he should be given attention, the freedom to run free, to sniff, explore, play and socialise.

THINK PREVENTION – NOT PUNISHMENT!

Some behavioural problems cannot be solved by purely positive training, and in these situations preventive strategies are required. A dog who chases sheep is not interested in the sausage or tennis ball he will get if he returns to his guardian. Chasing sheep is far more rewarding. Problems of this kind are avoidable if a dog is given appropriate training – setting boundaries and establishing clear communication – before his behaviour escalates.

I believe that many behavioural problems could be solved, or even avoided, if we spent more time with our dogs, trying to understand their behaviour and making sure their mental and physical needs are met. This means providing walks that last longer than just the few minutes required for toileting. Dogs need the opportunity explore their surroundings by sight, and most particularly by smell. In addition, important life skills, such as a reliable recall and a duration 'sit', should be under constant review, as this gives us the ability to manage distractions.

Many guardians struggle to analyse their dogs' behaviour and to understand what is going on. There is a tendency to think that the dog is doing something *"to annoy us"*, and we take his behaviour personally. Our emotional response clouds our judgement and we fail to identify where the dog's reinforcer lies, i.e., why a behaviour is worthwhile for him. This is not always straightforward and finding the right sort of help is not always easy. My greatest wish is that dog people would judge each other less and accept that the world cannot be divided into *only* positive or *only* negative. We need to listen to each other, be open to new ideas, and find appropriate solutions.

Our emotional response to a dog's misbehaviour often obscures our objective view of where the reinforcer lies.

So, what can we do when a problematic behaviour has already become established? What happens to the dog who is 'guarding' his toys/trophies and bites visitors if they come too close? This behaviour will not stop if we remove his toys so, as a result, such dogs often end up in rescue. There may be attempts to retrain the dog, or to find a new home without cats, sheep, visitors, etc., but if that is not possible, he may be euthanised. A learning process will only be possible if there are both positive and negative consequences, which means we cannot choose between purely positive training or punishment-based training. Instead, the discussion should be about when, how, where, and why we apply positive and negative consequences in our training:

- What is the desired or undesired behaviour that we want to strengthen or weaken?
- How can we work out exactly what we want to change?
- What will be the result of our positive or negative reinforcement and how can we measure it?
- What preventative measures could we impose instead of using punishment?

If a dog is chasing sheep, we cannot reward or punish him *at that moment*, except via an electric collar. Leashing would be a form of management but not a solution to the problem. Positive trainers would advocate that we start focus training at a distance where the dog is still able to respond. We can then reinforce – click and reward – for looking away from the sheep. This process would continue until the dog opts to ignore the sheep, even when they are in close proximity. So, there are positive training opportunities here. However, we should be aware that this form of training is not entirely positive; the dog may perceive the fact that he is not allowed to chase sheep as punishment. It is possible to teach strategies to behave well around sheep without negative expectations. This involves starting training from an early age on small animals and then generalising, moving on to larger animals. The aim is to train the dog to work in all environments, around all sort of animals, including sheep. For this, we need time, the correct strategy and, importantly, we should never allow frustration kick in.

For more information see Chapter 17: We need to talk about punishment.

FOOD ADDICTS

We love positive reinforcement and, although we are convinced of its efficacy, we still need to keep an open mind and try to be objective. For example, when I am walking dogs in my neighbourhood, I often meet an elderly lady who always has a plastic bag full of dog food with her. Every few metres she gives her dog a treat, because otherwise he will sit down and not go any further. I have even seen trainers giving copious amounts of food for almost every move a dog makes. The intention is to keep the dog motivated and interested, but the result is that the dog becomes a food junkie. It is no longer about behaviour training, it's about bribing the dog to move.

It helps to be aware of the difference between a reward and a bribe:

A reward is given ***because*** the dog has done something; a bribe is given so that he does something.

Inevitably, a dog who receives bribes becomes entirely dependent on the food reward. This becomes a major issue, as not giving addicts what they need causes tremendous stress and, in some cases, the results are very punishing.

Working with Bob Bailey and Marian Breland Bailey, I gained an in-depth understanding of operant conditioning, and how to apply it, using chickens as 'models'. We worked with a training schedule, taking small steps as we recorded and evaluated data. Sometimes we elicited a behaviour, stringing different behaviours together in a chain, and only reinforcing when the last link in the chain was performed correctly. These days, many of the chicken training camps or chicken workshops take massive shortcuts. I have seen students luring chickens with cups of feed, passing the cup over obstacles or through weave poles for the chicken to follow. This is not an effective method of training, even if the chicken completes a chain of behaviours. We must also be able to change an individual's behaviour without creating a food junkie who will do anything to get the 'drug' on offer. We need to train at a level where the animal is able to think and make decisions. The decision-making process can be slowed down, disrupted or even blocked not only by punishment but also by addiction-related stress symptoms.

This also applies to the use of the bridge signal. Let's take the clicker as an example. If a dog is clicked and fed all the time for everyday things, and not just when he is learning something new, we are in danger of creating an addictive behaviour. The dog is constantly on the alert, waiting for the next click. Mindlessly, he tries out different behaviours, hoping that something will earn a click. This is not simply because the click means food; the effect of classical conditioning is that the sound of the clicker leads to a release of dopamine. Now the dog feels great – and he wants more of it! This would not be a bad thing in itself. However, the knock-on effect is that the dog's state of excitement increases to such an extent that learning is no longer possible.

Incorrectly understood 'continuous clicking' can lead to real symptoms of addiction, and learning becomes a challenge in this heightened state of over-arousal.

ON/OFF SWITCH

We cannot switch learning on and off – dogs are always learning, be it unconsciously (classical conditioning) or consciously (operant conditioning). However, through training, we can add an on/off switch to start or stop a particular behaviour. A scent detection dog who is continuously sniffing will be tired out by the time he finally arrives at his work location. A protection dog who wants to bite everyone he meets will be exhausted by the end of the night patrol. For this reason, we attach specific signals or cues to specific behaviours. A well-conditioned signal can both start and stop a behaviour and can also relax the dog. Compare it to a torch that is left on all the time, just like a scent detection dog who is constantly sniffing, the batteries will run out. However, if we use the on/off switch on the torch, we can ensure the batteries still have sufficient power when we need them. We only switch on the torch when it's dark and we need to check something. Then we turn it off and save the energy for the next task.

Chapter 13:

OVERCOMING PROBLEMS IN THE LEARNING PROCESS

In the world of professional dog training, problems encountered during the learning process have traditionally been 'solved' through coercion. This is a sad fact, evoking both shame and guilt. But today we know better. We have an understanding of what slows down the learning process, and we test our dogs for suitability before training commences. This not only applies to sensory performance, such as hearing and sight, we have also developed simple test procedures to check nosework, problem-solving ability, as well as physical and mental capabilities. We also consider breed and personality differences.

I have been fortunate to be present at aptitude tests for potential recruits for the Norwegian and Swedish army and police. The dogs are tested at the age of 18 months to assess whether they have the mental strength and stability, and the talent, for this demanding work. Hundreds of dogs are tested in Scandinavia every year, accumulating an enormous volume of

A K9 aptitude test in Norway.

data and experience. Working in a wooded area, they are given multiple tasks and exposed to a wide range of distractions. They are tasked to search for items, they are approached by strangers, they hear gunshot, and are then invited to accept treats or play, and they are suddenly confronted by dummy characters that appear out of nowhere. The test is the same for all dogs and their performance is evaluated by experienced observers.

Testing in this way allows us to recruit dogs who are most suited to the desired role, often bypassing problems that might be experienced by dogs who do not show a natural aptitude for the work.

However, in the course of training, there will be issues, or obstacles to learning, that require our careful consideration.

UNINTENTIONAL INFLUENCE

The Clever Hans effect

Clever Hans was a horse who lived in Berlin in the years leading up to World War One, and his mental prowess took the world of psychology by storm. The Clever Hans effect, named after him, still influences animal training today. It refers to unintentional influence on the part of the trainer, highlighting that even the most subtle forms of body language can trigger reactions in an animal.

The horse's owner, Wilhelm von Osten, was convinced that Clever Hans was just as intelligent as a human being and could think, read and do arithmetic. First, he taught his horse to 'read and write'. Each letter of the alphabet was assigned a number and Hans was taught to tap the appropriate number with a front hoof. The training lacked sophistication – von Osten simply grabbed a front leg and demonstrated the movement, but over time Hans learnt what to do.

On June 28, 1902, an advertisement appeared in the *Military Weekly Gazette* as follows:

I want to sell my beautiful, 7-year-old, gentle stallion, with whom I am doing tests to determine the mental ability of the horse. He distinguishes ten colours, reads, knows the four calculation types, etc. van Osten, Berlin, Griebenowstraße 10.

In fact, von Osten had no intention of selling his horse, his plan was to arouse public interest. Initially this backfired as people thought it was a belated April Fool's prank, but after placing a second advertisement von Osten invited a crowd of people to his yard. There Clever Hans performed his amazing feats, which included telling the date and the time, and doing a variety of sums. His performance had a huge impact, and soon he was making headlines all over Germany. Von Osten did not charge visitors for admission money, but many still believed he was engaged in a money-making scam. In a bid to find out what was really happening, a group of scientists conducted a detailed investigation. After careful observation, they concluded that there was no sleight of hand.

It was the psychologist, Oskar Pfungst, who finally discovered the facts. He noticed that Hans was paying very close attention to von Osten while he was 'answering' questions by tapping his hooves. If he was asked for the answer to a sum – 2 x 2, for example – he would tap his hoof four times and then stop. Pfungst excluded von Osten and asked the horse a few questions himself. Again, Hans answered correctly! By this stage, most people were convinced that the horse was really able to count, but Pfungst still had his doubts. The next day he conducted another series of tests, but this time no one present knew the answers to his questions. Now Hans could no longer answer correctly. Pfungst's explanation was that Hans could read the smallest changes in body language not only in von Osten, but also in other people around him. His motivation for paying such close attention to these signals was that he was given bread and carrots for correct answers. Pfungst discovered that Hans was focusing on the questioner - paying attention to the tiniest, unconscious forward movement of the body or tilt of the head - and taking this as the signal to stop tapping his hoof.

APPLYING THE CLEVER HANS EFFECT TO DOG TRAINING

Animals, certainly horses and dogs, perceive the most subtle of body signals, even those that other people miss. This means we give signals to our dogs without even realising it. We can test this by asking for a 'sit' or a 'down' without using words. Without a verbal prompt, the dog will respond correctly. Better still, film someone practising 'sit' and 'down' with their dog and watch the footage in slow motion afterwards. Focus on the trainer's head and shoulders; they will move forward more when asking for the 'down' than when asking for the 'sit'.

Regardless of our acting skills, we cannot fool our dogs! This has implications in all aspects of training, but particularly in relation to scent detection. The handler can do a great job pretending they don't know where the search item is hidden but, all the while, the dog is busy picking up unintended clues. He only needs to look at his handler to know whether or not to alert, which involves no thinking, or decision-making on his part. In addition, he may also lose faith in his handler. Why should he use his nose and search a series of rooms or vehicles if he can detect from his handler's body language whether these places are worth investigating? When a dog understands the difference between training (handler knows) and working (handler doesn't know), problems will ensue.

The Clever Hans effect does not only take place between dog and handler, it also generalises to observers standing nearby. Pay close attention to the spectators during a scent detection dog trial: when the dog gets close to the hidden search object, people stop talking, move differently and even breathe differently. If we humans can perceive this, then we can be absolutely sure that our dogs are absorbing this information. According to the same principle, a tracking dog will pick up clues from the tracklayer, who walks behind dog and handler, issuing instructions as they work the track.

FLOODING

This learning technique is used as a means of overcoming fear. The animal is 'flooded' with the aversive until it no longer reacts to it, and is no longer fearful. It is an extreme way of putting an animal in a situation where he is maximally exposed to a stimulus and cannot escape from it. After an initial shock, the animal quickly realises that nothing bad is happening. For example a horse who is afraid of flapping objects would be securely tethered and

WORKING A DOUBLE BLIND

To prevent the dog picking up unintended signals in scent detection work and tracking, we need to work double blind, i.e., neither the dog, nor the handler, has prior knowledge. As early as 1947, the Swedish psychologist, David Katz, made the point that tracking training can only be useful and effective if the handler does not know the course of the track. Tracking on known tracks is worthless.

I am often asked when we should start double blind work. My answer is: as early as possible! Of course, we can't start it right at the beginning as, initially, we need to learn how to communicate by observing the dog's body language:

- What does he do when he encounters an interesting scent?
- What happens if he encounters a distraction?
- How can we detect if he is motivated or not?

The best way to do this is for the handler to lay the track or hide the target scent. Then, dog and handler should be filmed in slow motion. This gives the handler the opportunity to replay the footage, observe the dog's behaviour and analyse it. This is not just about the way the dog uses his nose, it is also the position of his ears, the movement of his tail, and his gait. Once the handler has some understanding of the dog's behaviour when following a scent, we can introduce double blind work. In this scenario, a tracklayer/assistant will lay the track or hide the target scent. When the handler and dog start work, the tracklayer/assistant will be on hand to coach, and give directions, if the handler fails to pick up on the clues his dog is giving.

This training process involves getting feedback and collecting data, enabling the dog and handler team to slowly work towards a success rate of around 80 per cent. Evaluation of work in progress is essential, which might entail a search in a field or searching a car ten times in a row. If the dog and handler are successful eight out of ten times, without assistance from the tracklayer/assistant, they are ready to start double blind searches. At this point, both the dog and the handler are literally working blind; neither know the direction of the track, nor where the target scent is hidden – and the tracklayer/assistant is no longer present either!

I believe it is absolutely essential for operational teams to be able to work double blinds, but it tends to be the exception rather than the rule. I have attended hundreds of training sessions worldwide, and I have seen some amazing work, but handlers always become nervous when double blinds are mentioned. I am always delighted when a handler asks me to set a track or hide a target scent – that shows real confidence! Training in this way is the best way of preparing for the challenge of working in competitions, and in the real world.

then touched with a crackling plastic bag until he stopped responding.

When used by an experienced behaviourist, this technique can be a quick way of solving a variety of problems. However, when used incorrectly, it can do so much damage that many trainers reject it. For example, the tethered horse might panic and, in his attempt to escape, he might fall or seriously injure himself – resulting in the opposite of what was intended. There is also the danger that the animal might fall into learned helplessness (see *page 135*). However, when used by an experienced professional, flooding can be an effective technique for resolving a dog's initial trepidation when exposed to a new experience. (see *page 132).*

EXTINCTION

I have previously referred to extinction, which involves weakening a previously reinforced behaviour until it disappears, as one of the key elements of training (see *Chapter 9).* I believe it is a good and, above all, a safe way to change unwanted behaviour, although it could be argued that we should train in such a way that eliminating unwanted behaviour is not required.

To be effective extinction training requires a well-planned protocol, which correctly identifies what is reinforcing the unwanted behaviour, patience and good observation skills. Consider the following scenario:

A dog who jumps up at visitors will sooner or later get attention, be it positive or negative. Attention is the dog's reinforcer. The key element is identifying the reinforcer that is keeping the unwanted behaviour alive. When this is removed, the behaviour will diminish and eventually disappear. At the beginning of the extinguishing process, the dog may become frustrated, fearful or even aggressive. When animals show this type of response, we know the extinction process is underway. If we keep calm – and keep at it – the behaviour will gradually weaken. It is also important to be aware of the likelihood of an extinction burst (see *page 79)* where the behaviour we are seeking to eliminate gets worse before we see an improvement.

FEAR AND PAIN

Some trainers work with fear and pain to change behaviour, making use of both negative reinforcement and positive punishment. However, it is rarely the case that fear and pain is inflicted for no reason. For some behavioral problems, it can be the last resort to change dangerous behaviour and save the animal from being injured, killed, euthanised, or to protect other animals and people.

Many trainers find it difficult to remain objective about the use of these techniques, so it is important to understand what they entail. If a dog is about to steal a sandwich from the table, a verbal reprimand, *"ah-ah!"* or *"no"*, and a change in body language are warning signs, telling the dog not to steal the sandwich. This can trigger a fear response. The question of whether we call the verbal reprimand negative reinforcement or positive punishment comes down to what we want to achieve as trainers:

- Do we want to punish the dog for stealing, or teach the dog to ignore distractions (like the sandwich)?

This shows how positive punishment and negative reinforcement are connected. Negative reinforcement cannot be used if the animal has never experienced positive punishment. The difference between the two is the goal.

- Do we want more or stronger behaviour?

If this is the case, we can work with negative reinforcement.

- Do we want to stop or get rid of a behaviour?

In this case, positive punishment is a possible technique.

In this example it doesn't matter what we call the technique. However, we do need to understand that what we are doing can create fear in the dog and may do lasting damage to our relationship. *For more information, see Chapter 17.*

If techniques that involve fear and pain are the only solution, then this must be clarified in a behaviour analysis, and should only be carried out by experienced professionals.

Fear and pain generalise quickly and the flight or fight response is very strong. These powerful emotions can be triggered by context, environment, experience and breed-specific dispositions, but distance is also a criterion. If a threat is close at hand, a dog will either run away or fight. The change in behaviour takes place quickly and it is not always possible to predict which option he will choose. This is often seen in chance meetings in the park; dogs may seem to get along at first, but then a scuffle breaks out if they get too close.

The latest research tells us that the flight or fight model needs to be updated. In scientific circles it is known as the 4-F model, incorporating freeze and fooling around/fiddle/flirt. When a dog freezes, he stills and becomes completely withdrawn. This may be the prelude to another fear response, such as flight. The fool around/flirt response is a means of releasing stress and can take many different forms of 'silly' behaviour. For example, a Cocker Spaniel will go into frantic search mode, but he is not following a scent, he is using the activity to relive tension.

Remember, a good, positive relationship between dog and guardian/trainer will always counter condition against fear.

LEARNED HELPLESSNESS

In the late 1960s/early 1970s, psychologists, Martin Seligman and Steven Maier, conducted a series of experiments that led to the theory of learned helplessness. We can observe this phenomenon in both humans and animals when they have been conditioned to expect suffering, pain or stress in a hopeless situation. The experiments initially focused on avoidance behaviour in dogs and rats. Today such experiments would no longer be ethical, but they were allowed back then and we have learned a lot from them. Seligman is considered to be one of the founders of positive psychology, which focuses on strategies to improve wellbeing. His work has given us deep insights into the human psyche.

Let's take a look at one of his experimental set-ups:

- A dog was placed in a box divided into two areas - A and B - by a small, flat partition in the middle. When the dog sat in area A, a sound would be played and a few seconds later the floor of the box would emit electric shocks. The dog jumped over the small partition into area B, where the floor was not electrified. The dog *fled*.

- Following a few repetitions, the dog was jumping into area B as soon as the sound was played and before the shocks started. The dog *was avoiding*.
- In the next phase of the experiment, the dog was placed back in area A, but the small partition was now replaced with an impassable barrier. The sound was played and the dog wanted to flee, but this was longer possible.

This is where learned helplessness begins; neither avoidance nor flight is an option. When animals (or humans) realise that they have no control over their situation, they feel helpless and act accordingly. Now comes the most interesting part:

- Following a few repetitions, the impassable barrier was replaced by the original, flat partition. As before, the sound was played followed by the electric shocks, but this time the dog remained motionless in area A, even though he could have escaped.

Learned helplessness can be observed in many situations – I sometimes see it when large organisations hold meetings. Once a person believes that they have no control over their situation, they become very passive.

Many training techniques involve avoidance in some form. This is a natural response when it comes to environmental influences such as heat, cold, etc. But an element of learned helplessness can also come into play, and we need to understand when and how this happens. Once again, we can see a connection with the flooding technique (see *page 132*); in fact, the line

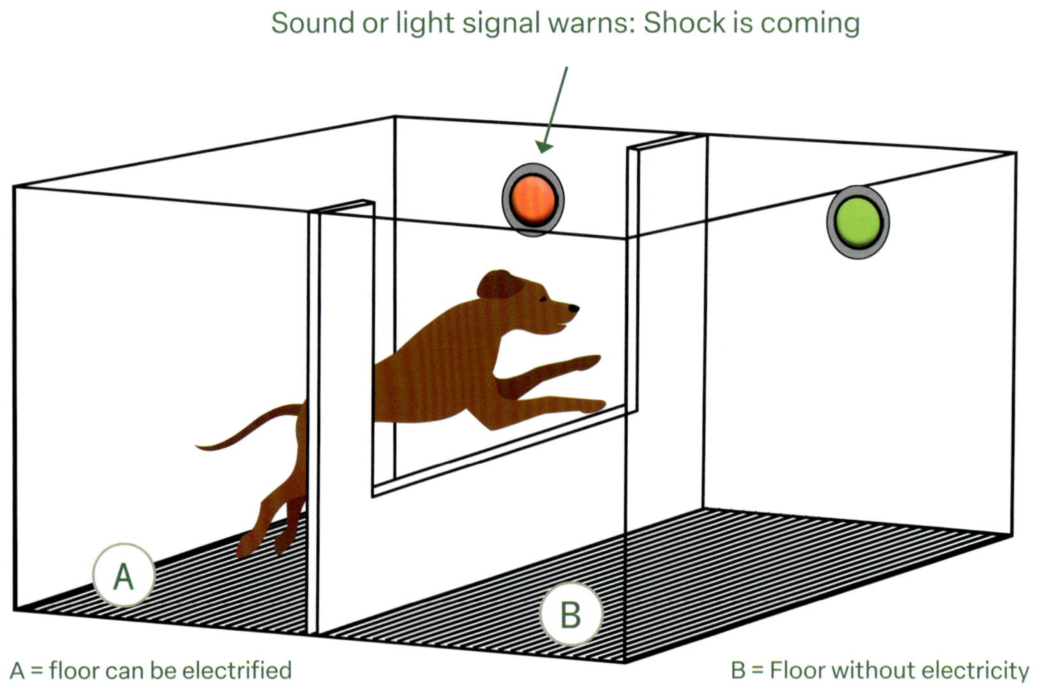

Seligman's experiment for learned helplessness.

between the two is very thin. If flooding is not carried out correctly, the result can be a state of learned helplessness.

For example, if a dog is afraid of loud noises we could put him in a closed box and expose him to loud noises until he stops responding. However, the dog has not learnt that the noises are harmless, he has lapsed into learned helplessness. Now we have created a far more serious problem – we have lost the dog's trust and ruined our relationship with him.

LEARNING IS ON 24/7

Never forget, learning is always happening – 24 hours a day, seven days a week! There is no on/off switch for learning. Dogs learn when they are out, and when they are at home; they learn both desirable and undesirable behaviours. It is, therefore, important that we, as trainers, are consistent with our dogs at all times, and in all settings. This is not easy; we are human beings and, unlike machines and algorithms, we make mistakes.

To establish consistency, we need to establish clear communication, and be aware of our body language, our tone of voice, and the context in which we are operating. For example, collar and short leash means going for a walk, long leash and harness means search work.

"Whenever you are with your dog, one of you is training the other."
Susan Garrett & Steve White

PUNISHMENT WEAKENS GENERALISATION

Thanks to learning by generalisation, our dogs are able to draw conclusions from differing objects and situations. They don't have have to try everything individually. Let's say the dog is learning to jump on to a raised platform. We practise this in the garden first and then, when we ask him to repeat the same behaviour on a slightly different platform, in a different environment, he knows what to do. He has generalised between platform and environment. We also see this process in target training. We first practise with the target on a wall and the dog quickly generalises the behaviour and touches the target when it is stuck on a door or on a car. However, it is important to vary the environment and context early in the training process. We should not continue practising until we are 100 per cent comfortable with a new behaviour before changing locations. We should begin to vary the environment and context when the dog is about 80 per cent fluent in the new behaviour.

Charlie generalises the platform, responding with the correct behaviour even though he is in a new environment, and using a slightly different platform.

But there is a catch: using penalty-based training techniques breaks the generalisation process – it slows it down or even destroys it completely. This is because punishment is generalised very quickly, much faster than positive reinforcement. We are not always aware of the side effects of punishment and when we identify them, it is often too late as it has been generalised much further than that one specific moment. This is yet one more reason to use positive reinforcement whenever possible.

Penalty-based techniques weaken or destroy the process of generalising newly learnt behaviours to other environments/situations/contexts.

THE LIGHT SWITCH PHENOMENON

Imagine coming home late and it's already dark when you open the front door. The first thing you do is feel for the light switch as you enter the hallway. In my house we have five switches right next to each other: for the kitchen, living room, hallway, first floor, second floor. Even after years of living in the same house, I still don't know which switch to use for each particular room. I just keep hitting switches until the hall light comes on. If the switches were located in different places, I would have worked out which was which to save myself the labour of going to each switch in turn until I got the result I wanted. Spending too much energy on a task is not economical and will lead to a change in behaviour. This is where so-called behavioural economics comes into play – and it also plays an important role in animal training.

If a detection dog learns to sit when he identifies a target scent, then his sit will become faster and faster as it is associated with reinforcement. After a while, the dog will give a very quick sit if he finds an odour similar to the target scent. The sit is a simple behaviour and the dog does not need to expend much energy to do it. Therefore, a false indication is a simple, low-cost mistake. However, if we ask the dog to do more – standing on his hindlegs and pinpointing the target scent on a wall – he will burn more energy. Mistakes will be more costly, and he will, therefore, change his behaviour more quickly.

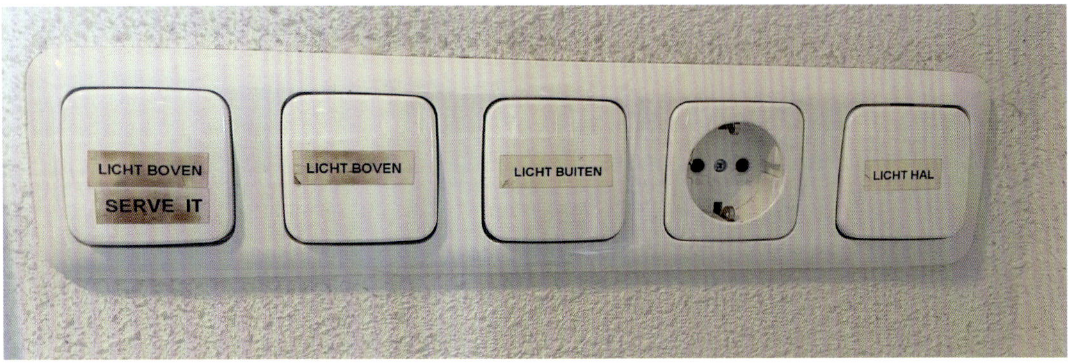

Now the light switches are labeled – because of their close proximity, I never learnt which was which.

Chapter 14:

MOTIVATION AND STIMULATION

If a dog – or any animal – is to behave as we prescribe, he needs to be motivated. That is, he needs to see why he should co-operate. There are two types of motivation: positive and negative:

- If I want you to put on a winter coat, I can take you to a very cold environment. Putting on the warm coat is *positive* for you.
- If I want you to take off your coat, I can heat the room you are in. It could be *negative* for you to keep wearing the coat.
- Imagine I ask you to open a door. Behind that door there is £100 which you can keep. Now you have a *positive* motivation for opening the door.
- Or I could point a fire-hose at you and tell you that if you don't open the door, I'll open the valve. Here I'm using n*egative* motivation to make sure you open the door.

But there's a catch: What if you have just won a £1million on the lottery? Would you still open the door for £100? What if I aim the fire-hose at you on a hot day and you welcome the chance to cool off? Motivation and response are influenced by the environment, distractions, and the individual.

B.F. Skinner's exposition of operant conditioning is based on positive and negative motivation. In his book, *The Behavior of Organisms*, he describes numerous experiments in which he used these principles to modify and condition behaviour. When working with operant conditioning, it is important for us to know what our trainee perceives as positive or negative. In fact, this is equally true of classical conditioning. We need to be able to answer the question*: 'what's in it for me?'* from the perspective of the learner or the animal being trained. Although classical conditioning is more about reflex behaviour, it is underpinned by motivation:

- If your skin feels a heat stimulus, you retreat to avoid burning yourself.
- If something comes close to your eye, you close it to protect it.
- If you are hungry or thirsty, you will eat or drink.

Therefore, before you write your training plan and start training, you need a full understanding of what motivates your dog.

In addition to positive and negative motivation, there is a third motivator that changes behaviour. Known as intrinsic motivation, it is a motivation that comes from within the individual. It is the persistent striving to get better at something or to achieve what others think is impossible. The incentive is less the achievement of a specific goal than the performance of the activity itself. To better understand intrinsic motivation, take a look at the following examples of individuals who kept on striving:

- Walt Disney was fired for lack of imagination.
- Michael Jordan retreated to his bedroom in tears because he thought he would never learn to play basketball.
- The Beatles were rejected by several

producers telling them that no one would like their sound.

We can also use the example of toddlers learning to walk. They fall frequently and yet they keep trying.

Sometimes it's difficult to answer the question: '*what's in it for me?*'

- Why do some people keep getting up and trying again when everyone else thinks they have failed?
- Why do some people always want to learn more, and train harder, and are not easily satisfied with what they have achieved?

It's all about previous experiences and intrinsic motivation. When I trained as a military diver at the age of 17, I learnt how to survive in the most difficult and dangerous circumstances. That had a major impact on me and has helped me ever since in all kinds of different situations. But it can go back much further. For example, did your parents gave you enough attention when you were growing up? Was there rivalry and your brothers and sisters?

MANIPULATING THE ENVIRONMENT

Trainers are masters at manipulating the environment. This is important because it helps set animals on the path to success. I like to start basic training in a small room without distractions - no furniture, nothing on the walls - just the trainer, the dog, and the training equipment, such as a target or a scent wheel. As training progresses, I gradually introduce distractions. Then when we go outside, I make sure that I can control the environment as much, and for as long, as possible. This is the way we prepare animals for working in the real world, where we can no longer control the environment.

As the tasks become more difficult, involving increased duration and distance, etc., I also ensure that the dog is successful most of the time. If we allow animals to continue working, despite numerous failures, they will eventually give up. For this reason, it is important to collect data and control the training process.

In the real world, a dog has to focus and work, regardless of distractions

MENTAL STIMULATION

This involves engaging with our dogs in everyday life so they are kept mentally alert and busy, but without stress. It's about being part of the family routine, with added challenges that provide mental stimulation. I focus on physical aspects, such as running, jumping, negotiating an obstacle course, retrieving, crawling, slalom running, swimming, and much more. In addition, I make use of dog intelligence toys to encourage problem-solving skills, and I also set up scent discrimination tasks which are mentally stimulating, rewarding, and give a sense of wellbeing. As the dog becomes proficient, it is important to introduce distance and distractions to add to the challenge. For example, a scent wheel with only one target scent is not particularly taxing, but when it is filled with distractions such as cheese, chicken or other odours that closely resemble the target scent, it's a real workout for the brain!

AROUSAL LEVELS

As previously discussed, it is important to find reinforcers that a dog truly values (*see Chart toppers, page 115*), but it is equally important to suit the reinforcer to the required task. When I teach a dog something new, I generally use food as it is a low-drive reinforcer associated with little excitement. The food is easy to handle, and the dog can swallow it quickly, so training can be resumed without delay. This means I can do a series of reps in a short amount of time. In contrast, a highly charged or high drive reinforcer, such as playing with a ball or a tug, demands a lot more time. In addition to the time factor, new skills learnt in a low state of arousal are better remembered.

The advantage of a high drive reinforcer is that it can be used to reward a particular challenge, such as completing the required task despite distractions. For example, a police dog in training might get a longer 'mock fight' with the 'criminal' if he has coped with a high level of distraction, such as fireworks. Or a dog may be rewarded with a great game of interactive tug if he has completed an obedience exercise in a field full of rabbits.

Obviously, it is important to work out what reinforcers are perceived as high or low drive, and this depends on the individual dog. So, in addition, to finding out what he values, we need to calculate the effect they have on his arousal level and, therefore, on his ability to learn.

MOTIVATION TO KEEP GOING

American coach, Tony Robbins, works on the basis that we, as humans, avoid pain and seek pleasure. The same also applies to animals. It's a clear and simple description of what drives us. However, some coaches also advocate a degree of suffering. If we want to bring a dog to the level of a well-behaved and obedient animal, we must invest a lot of time in training; in other words, we have to suffer! If we don't put in the necessary training time, we will end up with an untrained dog and both guardian/trainer and dog will suffer even more...

This interesting perspective makes it clear that, to make progress, we cannot remain relaxed and unchallenged. We all need to put our words into action, burn energy and commit to it. If we are willing to suffer, we can achieve anything we want.

Even with the best dog in the world, we still need to invest time and energy in training. Motivation, and a certain willingness to work, are elements that we wish for in dogs, but also in trainers! Too often the focus is on the dog alone and we seem to forget that the trainer also plays a role. Malcolm Gladwell, an American writer, popularised the so-called '10,000 hour rule' which states that if we want to reach a level of excellence, we have to put in at least 10,000 hours of training. I recalculated this using the model of an active police dog handler:

Imagine a police dog handler has one official training day per week, estimated at eight hours. When he arrives at the training ground at 9 am he meets the other handlers and they have a coffee together. After half an hour, the chief trainer explains the programme for the day. That takes another 30 minutes, and already a whole hour is used up. Now it's time to start the training, first assembling the equipment and letting the dog relieve himself. Another 30 minutes are gone. It's 10.30 when the training finally begins. There is a total of 10 handlers and dogs, and they have to work individually. After an hour and a half, it's lunch time, and our police dog handler has worked with his dog for 15 minutes. After the one-hour lunch break, he is lucky and can work with his dog for an hour and 45 minutes in the afternoon training session. Of the planned eight-hour training day, he effectively had almost two hours of training time. If he can train for the same amount of time every week, he may train two hours a day for 50 weeks, making 100 hours a year. So, before he reaches the magic 10,000 hours, he has to work for 100 years. What if our police dog handler is a real fanatic and has spent almost all the scheduled time training his dog? He goes without coffee breaks and lunch breaks and manages to train seven hours a week. Even so, he still needs 28.5 years to reach 10,000 hours.

Full-time trainers working 40-hour weeks have to organise classes, manage their business, and deal with other distractions. On average, I have calculated that they will effectively work 20 hours per week for 50 weeks per year. That makes 1,000 hours of training per year, taking 10 years to reach 10,000 hours. Of course, Malcolm's 10,000-hour rule is not set in stone, but it gives us a good guide as to how much investment it takes to reach a reasonable level of expertise. To achieve this, we need to approach training with the right motivation: trainers need perseverance!

As with dogs, human motivation can be altered by using positive and

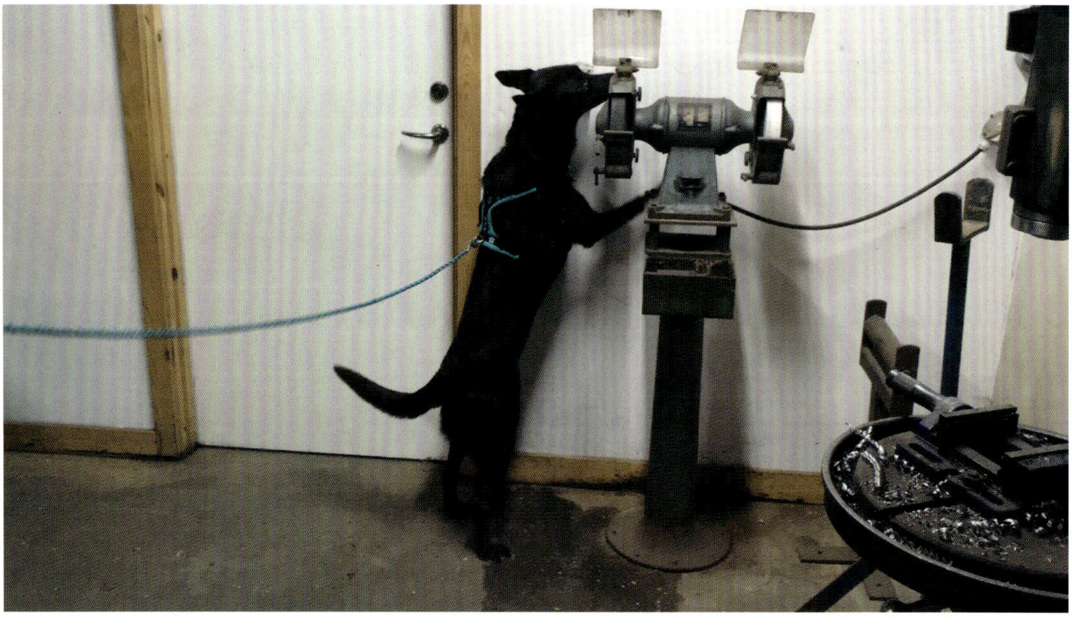

When searching for difficult objects, motivation is especially important.

negative reinforcers. Just like dogs, we humans prefer positive reinforcement. Here, however, lurks an interesting pitfall: children who have always been spoilt by their parents never learn to fight for anything. They just get whatever they want; they don't have to go to school when they pretend to feel sick, and they never hear the word 'no'. In contrast, the children who have had to fight for the things they really want and have been disappointed many times before reaching their goal, are much tougher and show more stamina.

History has its own examples: Abraham Lincoln suffered a series of defeats and setbacks before he was finally elected President of the United States. It's all a matter of perseverance and motivation, as well as picking the right people, and making them the best coaches. And even with the best training, we still need to keep the participants motivated.

NOSEWORK AS A MOTIVATOR

The power of nosework has been researched by Dr Charlotte Duranton and Dr Alexandra Horowitz, and they have come up with some fascinating results. They have found that dogs who do nosework show increased optimism over those who are only worked in traditional obedience exercises.

Both disciplines worked with food as a reward, but the dogs involved in nosework had more autonomy than in obedience, which appeared to be the significant factor. In their study, the two scientists highlighted the importance of providing choices, allowing dogs to make their own decisions, and giving them the opportunity to use their sense of smell, which improved both mood and general wellbeing.

A total of 20 adult dogs of different breeds took part in the study. They were divided into an obedience group and a nosework group. Before starting their training tasks (in nosework or obedience), the dogs from both groups learnt that a bowl located in a particular place always contained food, while another a bowl in a different location never contained food. Following training, the dogs from both groups were tested on the set-up but now there was an empty bowl in a new location, equidistant between the two bowls in their original places. The length of time a dog pondered over whether there might be food in the new bowl, i.e., his anticipation, was used as a measure of his optimism.

The results suggested that the 'nose group' was generally more optimistic in their expectations. One of the reasons could be that nosework training allowed these dogs to roam freely, to scout locations, and to solve problems - and successful problem-solving creates a feeling of wellbeing. Another reason for their improved mood was that they were allowed to use their noses. Smell is an important source of information for dogs, and the act of detecting a scent and processing it, enriches their everyday lives.

This fascinating study shows us, once again, how we can deal with situations in our everyday life. Is the glass half full or half empty? If we, as trainers, can motivate our dogs - not only through rewards, but also through expectations - we have a powerful tool which we should build into our training methodology.

Chapter 15:

WHAT MAKES A GOOD TRAINER?

You may have the most talented dog in your chosen discipline, but unless you have the right people involved in the training process, you will never achieve your full potential.

When I started my career as a young police officer eager to work with police dogs, the system was not built on achieving excellence. At that time, the general K9 police unit was mostly comprised of recalcitrant officers who either had problems with the hierarchy or were prone to violence on patrol duty. Most didn't even like the dogs they were assigned. They were more interested in the company car they were allowed to take home and the extra payment they received for keeping the dog. Most training days were spent playing darts, drinking coffee and gossiping.

Thankfully those days are over. K9 units have become professional outfits, and officers take a deep interest in the training process. They take good care of their dogs, spend a lot of time with them and give them the mental and physical stimulation they need. There is always room for improvement, particularly at management level, but the system is now unrecognisable from my early days in the force.

In a professional setting, such as the police, the military, or assistance dog training, there is a clear demarcation between the trainer and the handler. But the lines becomes blurred in the world of performance sports, where the top trainers are often the top handlers, and in the vast area of training and behaviour where trainers run classes and workshops to showcase their own methodology. For the time being, let's put the role of the handler to one side, and focus on what the trainer/instructor/teacher brings to the table:

- What is teaching?
- What can we expect?
- How do we recognise good and bad teaching?

A teacher's role is to transmit information. Good teachers make us curious to find out more – to find out what is possible. A good teacher presents us with challenges. leading the way and encouraging us to step out of our comfort zone. Such a teacher can make a tremendous difference to their students. Teaching and learning is not just about storing information in students' minds. More importantly, the information should be used when it is needed, thereby facilitating success in the operation, the competition, or the given situation or task.

Let us remind ourselves of the definition of training: it is nothing more than changing behaviour. When we work with animals, that's exactly what we are dealing with every single day. In addition, we reinforce, ignore, or punish behaviour. So, what makes a good trainer? The best trainers are able to change behaviour quickly. In addition:

- They are creative in altering environmental conditions.
- They use all sorts of reinforcers.
- They have a natural connection with animals.

- They are experts in timing and body language.
- They can observe behaviour in its finest detail and make quick decisions.
- They know what they are working towards and are focused.

For the majority, training is a motivator in itself; trainers see their mistakes as something to learn from. There is also a difference between trainers who enjoy working with animals and those who enjoy working with both animals *and* people. I have met a number of talented dog trainers who had absolutely no motivation to work with people, which rules out the possibility of passing on their skills. Those who set themselves up to train should be judged on their performance and their motivation. They must be able to produce reliable behaviour at speed, but they must also have the ability to communicate with people, working with their strengths and weaknesses, and their preferred method of learning.

ESSENTIAL SKILLS

Listing the skills required by a good trainer is an impossible task. The list can never be exhaustive and there will always be debate as to what is considered important/unimportant/essential. I have met many trainers I have respected over the years – young, old, novice and expert – and they all showed special and sometimes unique skills. Perhaps what stood out was that they would never give up. Like toddlers learning to walk, they kept falling, but they would always get up and try again. The toddler sees his older siblings running around effortlessly but is not demotivated. It is an inspiration to keep trying until they are just as good, or even better, than their siblings.

So, if I had to make a list of essential skills for a trainer, this would be my selection:

- **Persistence:** Keep going! Don't give up, even after many failures. Training is simple, but not easy.
- **Observation skills:** Good trainers seem to have a second pair of eyes, picking up on the smallest details of the dog's body language, and also taking note of the environment with all its possible distractions.
- **Self-reflection:** I rate this quality very highly. Those who cannot reflect on themselves will not learn from the mistakes they will inevitably make.
- **Willingness to change:** The involves not only the desire and ability to change the animal's behaviour, but also their own.
- **Vulnerability:** This is often perceived as a weakness, but I believe it is quite the opposite. It is the willingness to take emotional risks, to face a situation and to stand up for what we believe in. Vulnerability brings new ideas and creativity into our lives.
- **Record-keeping:** Collecting and analysing data is an increasingly important in animal training. It's the only reliable way to find out if we are making progress and if we are on the right path.
- **Broad interest in animal training:** A good animal trainer should not be limited to their specialist area. Dog trainers from different disciplines can learn from each other. Indeed, dog trainers can learn from trainers of birds, and many other animal species. We can also learn a lot from looking at how animals behave in the wild and learning from them.
- **Communication:** The ability to express ideas, and to give instructions is key. This

Decision making during a double-blind search.

goes hand in hand with listening, which allows us to learn from others.

- **Patience:** A good animal trainer must be able to wait for the specific behaviour they want to reinforce.
- **Confidence:** Not to be confused with patience and perseverance. Most of the time we let the animals decide if a trainer has confidence – we can tell by the way animals react to a new trainer. Confidence also means that the trainer sticks to their plan and takes the lead and does not get side-tracked by extraneous distractions.
- **Mechanical skills:** This means the trainer can implement plans on a practical/physical level. A trainer needs the control and the physical aptitude to communicate, to work with or without a leash, to be able to use tools, such as a clicker, and to deliver treats. I have been in the company of smart and intelligent animal behaviour experts who could give a full explanation of learning behaviour and training principles, but who did not have dexterity to train animals themselves.
- **Understanding of training principles:** This may seem obvious, but I have met quite a few trainers who knew nothing about the principles of classical and operant conditioning. These are the so-called 'gut feeling' trainers. Some of them do a wonderful job with the animals they are responsible for, but the scope is necessarily limited without theoretical knowledge.
- **Decision making:** A training process consists of making decisions. Should we reinforce? Help? Wait a little longer? What should we cue next? Are the approximations leading to the desired goal? What should we change? These, and other many other decisions, may need to be made in fractions of a second.

THE SELECTION PROCESS

Professional organisations, such as the police, the military, assistance dog charities and rehoming charities are involved in recruiting trainers on a regular basis. The process may not seem strictly relevant for independent trainers who run their own businesses, but I believe we can all learn a lot from objective assessment and detailed scrutiny.

Studies show that around 30 per cent of the selected candidates in all professions do not complete their training. Consequently, it is difficult, if not impossible, to come up with a 100 per cent reliable selection process. There will always be failures. The selection process is always difficult – not only with animals but also with humans. So how do we test our human trainer candidates, be it in police dog units, dog sport clubs, dog training schools, assistance dog and rehoming charities?

First and foremost, I want to see a candidate who is not only able, but also willing to do the job. Sounds obvious? But I well remember interviewing a guy, and his first question was about what company car he might get – and he was deadly serious. Needless to say, he didn't make it to the next round. I also remember an applicant who wanted to be a trainer for the new radio directional camera dogs. When we showed him video footage of a training session in which a trainer repeated a target behaviour about 700 times, he responded by saying: *"I don't have the patience for that – I want to work with a camera dog!"*. He didn't make the next round either.

Sometimes practical tests are more revealing than a formal interview. For example, I invite prospective trainers to run an agility course with pre-trained dogs. After the applicant has run the course with three different dogs, I give him a dog that has no previous training. I am always surprised by how quickly some of the candidates lose patience and resort to cajoling the dogs when they fail to clear the hurdles. Of course, what we are looking for are those who are creative and change their own behaviour to help the dog achieve the task.

I have also had some interesting results from a test that involves giving a trainer a young dog, a chair, a long leash, a clicker, and a bowl of food. The trainer must stay in the starting position – 5m from the chair – and teach the dog to jump on to the chair and sit on it for 30 seconds before being rewarded. The candidate has to write a training plan and collect data during the training process. This gives us the opportunity to see if the trainer sticks to their plan, or whether they keep changing their approach.

There is a value in interviews and in one-off practical tests, but I believe the best way to assess and select new trainers is to simply allow them to work with the dogs, and not just for one day, but ideally for a whole year. This gives us a real opportunity to see how the trainer treats the animals, if they are willing to put in the hard work, if they are learning, and if they are genuinely interested in training. After a while – when they have had the opportunity to observe other trainers – we ask them to practise some simple behaviours. We watch them work, and we are on hand to answer questions and to step in to help if required. I am not interested in their high school or college degrees, or how long they have been studying animal behaviour. Animal trainers must be able to apply training techniques and not just reproduce them theoretically in a test. I am also interested in the trainer's personality traits, their motivation, their ability to learn new things, and their observation and timing skills. Does the trainer have the ability to:

- Step out of their comfort zone?
- Reflect on what happens?
- Listen, and follow directions?
- Maintain a positive attitude.
- Make changes to their own behaviour, at speed.
- Read training protocols.
- Collect data.

On top of all that, I want to see how the dogs react to the trainer. To assess this, we work with a test that has been designed to measure the dog's level of comfort in the presence of a trainer. The test is simple: the candidate waits in a small and almost empty room. An assistant brings a dog into the room, unleashes him, walks out and closes the door. We observe the candidate and the dog from another room through a separating pane that is mirrored on one side, or by means of a camera. The task for the applicant is to play with the dog and arouse his interest. The applicant does not know this beforehand and therefore cannot bring toys or food for the dog. There are two indicator lights in the room: one green, and one red. The green means start playing, the red means stop the game.

As the test gets underway, we observe the candidate's behaviour. Does he/she:

- Wait for the green light to make contact with the dog

    or

- Start interacting as soon as the dog is in the room?
- Keep checking to see if the red light is on, desperately hoping for the session to end

    or

- Forget the time and the indicator lights and focus on playing with the dog?

After a few minutes, the assistant comes back, takes the dog away and brings in another dog. We repeat this a few times with different dogs; we focus on observing how the dogs react to the candidate and what he/she is doing to keep the dogs engaged. We then ask the applicant to train a new and unusual task within a fixed time limit of 20 minutes. For example, a dog has to jump into a wheelbarrow and stay in it while it is pushed for a distance of 20 m. We watch from a distance, observing the candidate's approach, their ideas, whether they ask for help, whether they use food or force – yes, that has actually happened in a selection process.

We also use clicker games to get an idea of the candidate's timing and observation skills. We deliberately give our instructions in such a way that the applicants have to ask for further information in order to be able to progress with the task. But we also give them a task they cannot solve – and put some pressure on them – to see how they can handle criticism. In addition, we show a short video of a training session and ask them to write a training protocol for other trainers, which can be clearly understood. This process allows us to gain a comprehensive picture of the applicant, their technical skills, their motivation, their patience, and whether they can build up a rapport with the dogs they are training.

MOTIVATING TRAINERS

Even the most motivated trainer can enter a period of low motivation. The reasons are not always clear, although, with good data collection, it is easier to predict when these moments are likely to occur. The first signs are the trainer puts in less effort and their success rate falls. At this point, supervisors should step in, not only to help the trainer, but also to make sure the animal is OK. Remember, we always get the behaviour that we reinforce, so an inconsistent trainer will train inconsistent behaviour. Training mistakes are not easy to correct, so we absolutely need trainers who are motivated and focused. Good trainers

Motivation curve

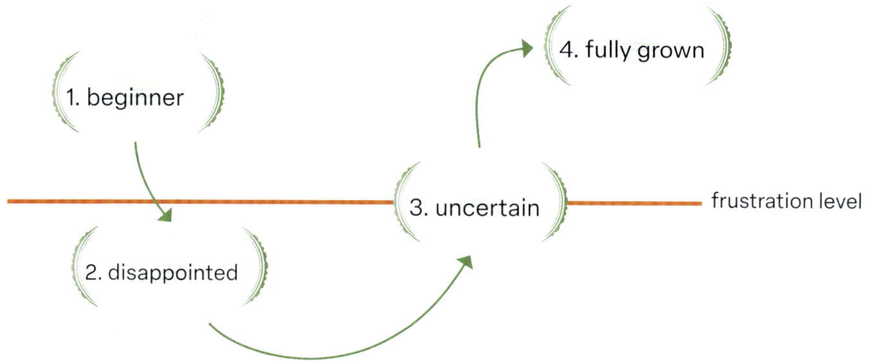

simply explained

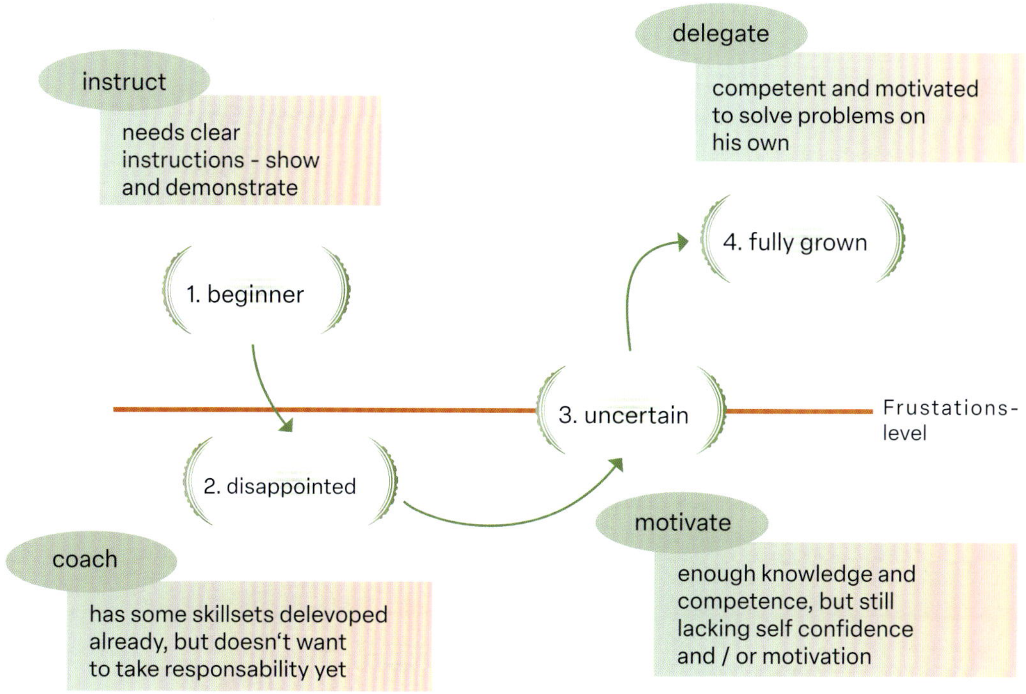

keep records and use various reinforcers to keep their animals motivated. But what do supervisors do to keep their trainers motivated?

In my experience, working with clear goals is a very good motivational tool. Some of our training programmes are scheduled over 12 months; motivating a trainer to start something and wait for 12 months before seeing the results is almost impossible. But by measuring the programme in a series of milestones, which can be reached every few weeks, we can build anticipation and motivation. In this way, trainers work from milestone to milestone, and whenever they reach a milestone, we check the results in a double-blind test. For example, we would set up a scent detection test where neither dog nor handler – nor anyone in the vicinity – has knowledge of the target scent. If a team can succeed in solving this double-blind puzzle, it provides a real boost to confidence and motivation.

When the human-canine team passes the test, we celebrate! After that, they can both move on to a new goal. However, they also need to preserve the skills they have acquired, and conditioned in their dogs, maintaining them at a level of excellence. This requires supervisors to regularly review performance which, in turn, allows trainers to show off their work and receive praise. In this way, everyone benefits: management and trainers know that tasks can be performed at the highest level, and the animals continue to be challenged, trained and rewarded. Throughout the training programme, we also keep an eye on the data; regular analysis ensures that everything stays on schedule and on budget.

I have met successful trainers who are very good at training dogs, but not at training other trainers. There are also trainers who are only good at training dogs of a specific breed or personality type. It is important for a trainer to be creative and to be able to solve all kinds of learning problems that arise – both in dogs and in humans – and to do this quickly and flexibly. It is important to remember that there are different personality and learner types.

There is no right or wrong or no 'better' learning style. In fact, we often see a mix of styles. However, what matters is that the trainer appreciates the fact that students learn in different ways, and it is their job to present information and instructions in the way that is best suited to the individual. Creating a safe learning environment and building trust are important tasks for the trainer; learning cannot take place unless participants feel secure (see *page 120*). Safety is not only about avoiding physical hazards, such as sharp, metal objects, chemicals, trip hazards, road traffic or other dangerous elements, mental safety also plays a very important role. We need to create an environment where people dare to ask questions, make mistakes and try new things. We want students to get out of their comfort zones, to show their weak side. Remember, asking for help enables an individual to grow to the next level.

Trainers should also keep checking that they are on the right path and that the students' learning curve has an upwards trajectory. Training is about asking questions, focus, and presenting information in such a way that students remember the lesson and can reproduce the behaviour in question.

TYPES OF LEARNING

Visual
Students learn from what they see

Kinesthetic
Students learn from what they do

Auditory
Students learn from what they hear

Stress
Students learn from what stresses them out

Ease
Students learn through what relaxes them

Scribble
Students learn by writing things down

Trust
Students learn from authority

Teaching
To pass on information, students must have a clear understanding of the concepts and practical skills involved

Copying
Students learn by copying

HOW TO RECOGNISE A GOOD TRAINER

There is a widespread belief that a good trainer will have the best behaved or most successful dogs. But is this true? Rather than judging a trainer on reputation, isn't it better to watch them work with their dog and see what they can do?

After all, everyone sets their own goals and rules. There is no universal law regarding a dog's ability to learn, or to perform set tasks. We have all seen working and performance dogs who have gaps in their education. For example, a scent detection dog be lacking in basic obedience, or an agility dog may run off with his toy. A good trainer is thorough and works through a training plan which has a beginning, a middle and an end.

An individual might reach a high level of excellence as a trainer, but that does not mean he can pass on his skills to others. There are people involved as well as animals, and a good trainer must be able to work with both. There is a saying that, for me, sums up the art of training: *"If you can't explain it simply enough, you haven't understood it well enough."*

A good trainer must be able to motivate, strengthen, encourage and develop not only the dogs (or other animals) in his care, but also the people.

A trainer should have the ability to bring out the best in everyone, recognising the needs and areas of concern of the individual and addressing these both individually and in a group. There may be times when this involves uncomfortable conversations, but these should never be avoided; the end goal is to keep everyone on board

A good trainer is like a captain – he can 'organise' the whole ship so that it reaches its destination.

When I am working with novice trainers or police dog handlers, I break down the basic skills into their key elements to promote learning and understanding. By giving them the opportunity to learn a wide range of manual skills and training methods, they acquire the basic building blocks to develop competence, confidence and motivation. This, in turn, encourages students to try new things – to be creative in their training. Most importantly, I reinforce the trainers, as well as the dogs. I work with video recordings, and I collect and analyse data to show the students the progress they are making.

Basic manual training skills are equally important. This might involve working with bridge signals, reinforcing close by or at distance, leash handling, body language, position, movement, use of keep-going signals, and so on. Once new trainers become confident in these skills, they can also learn other training-specific skills and complex movement patterns. A trainer who spends time on the basics, understanding the principles of training and mastering the practical skills, can look forward to a long and successful life working with animals, and will always be ready to face the challenges that lie ahead.

WORKING TOGETHER

When politicians are fighting an election, they have no intention of working together. In fact, they are at pains to point out each other flaws and the divisions between them. Unfortunately, bridges can never be built in this way. The same applies to the world of animal training; we will never make progress if we focus on the things that divide us. We need to bridge the gaps, not only between trainers, guardians, and scientists, but also between trainers. Animal training is our common interest, and we can all learn from each other. I am grateful to the people who showed me the path away from punishment-based training to positive reinforcement. Now I feel I should put myself in the shoes of others who have not yet had the opportunity to learn about the positive and negative side-effects of different training methods. So, I make this urgent appeal to the animal training world: *Let's work together instead of against each other!*

We need to stop polarising the argument and pointing fingers at others. Instead, we should explain, discuss and listen. We should help others instead of excluding those who think and act differently from us. We need to talk to *all* trainers, welcoming coercive and punishment-based trainers alongside positive trainers, the novice and the experienced, the old and the young, the traditional and the innovative, men and women, transgender people, gays and lesbians, atheists and trainers of all faiths. We can all learn from each other.

Chapter 16:

ADVANCED TRAINING TECHNIQUES

Now that we have looked at the basics of training, it's time to focus on the details which will become important as we attempt greater challenges. Obviously, this will involve more training, which begs the question: how frequently should we train, and how long should each training session last? There are many different opinions, experiences, and paradigms on this subject, but not much research. Key facts are missing, such as:

- How do we measure 'pure' training time, i.e., the time in which the animal is actually engaged?
- What constitutes training time? How do we distinguish between each element:
- Preparing the session
- Setting up the equipment
- Reviewing notes from the previous session
- Walking the dog prior to training
- Playing with him at the close of the session
- Writing down the data we have collected
- Cleaning up the training room
- Preparing for the next session, etc., etc.

To be able to calculate training time, I like to work with clear criteria - i.e., I calculate the trials of each session. A session begins when the dog and trainer start working together in the training environment. Within the session there will be a series of trials which, in turn, give us an exact training time. Let's take a session on training 'sit' as an example:

It's the eighth session on this exercise, so we will call it 'sit 8'. The session, 'sit 8' consists of 10 trials and each successful trial lasts 30 seconds. In session 8, the dog will sit for at least 300 seconds. The session, as a whole, lasts longer because the trainer is also occupied with reinforcing the dog, recording the data for each trial, and regaining the dog's attention before each new trial.

Dogs have their own personalities, interests, and motivations, and perhaps different attention spans, just as we see in humans. It is therefore difficult to give a precise calculation regarding the length of a training session. I have a rule of thumb: continue training for as long as the dog is engaged and interested in the process. I prefer short sessions, consisting of 10 trials per session, which the dog can complete in a matter of minutes. Then I take a break for a few minutes before moving on to the next session. If individual trials take longer, for example, I am increasing the criteria for distance, I might limit it to two or three trials per session.

It's also important to consider the end goal. Is it:

- Competing in a flyball competition?
- Executing a move in canine freestyle?
- Following a track of 1 km, or 10 km?
- Searching 25 cars for explosives or drugs?

With our intended goal in mind, we write our training protocol and work towards it until we reach the required level. There is an interesting study on this in *Applied Animal Behavior Science* No. 133 (2011) entitled *The effect of frequency and duration of training sessions on acquisition and long-term memory in dogs*. The study involved 44 Beagles who were divided into four groups and then trained in an obedience exercise. Each dog took part in 18 training sessions.

Group 1 trained once or twice a week and completed one session at a time.
Group 2 also trained once or twice a week but undertook three sessions in a row.
Group 3 trained daily in one session.
Group 4 also trained daily, but in three sessions in a row.

After four weeks of training, the results were as follows:

- The dogs in groups 1 and 2 (trained once or twice a week) performed better than those that had been trained every day (groups 3 and 4).
- There was no difference between one session a day (group 3) and training once or twice a week with three sessions in a row (group 2).

When assessing the trained exercise at the end of four weeks, it was found that the different schedules had no effect on the performance of the task. All dogs were able to complete the exercise, even after a four-week break. So once a dog has learned something, he has the ability to recall it for at least four weeks. This is a very interesting finding that could change the routine of dog schools and training centres.

I noticed something similar several years ago when I was involved in training two young explosives dogs. The training schedule was five days a week with about four to five sessions for search work per day, plus one or two for general obedience. However, we were pressed for time, so we decided to work with the dogs twice a week and give them one search a day. During the rest of the week, the dogs were cared for by staff members. To our surprise, we found that the dogs in question were ready for the exam at exactly the same time as those we had previously trained on the routine schedule of four to five sessions per day.

I am not suggesting that we should only train a dog twice a week. But the influence of Parkinson's law (see *page 31*) – the adage that work expands to fill the time allotted for its completion – can sometimes block innovative insights in training programmes.

We were using five days a week to train explosives dogs because that was what we had always done. However, when time constraints forced us to make a change, and we discovered we could train just as effectively in two days, we were able to use the other three days to train other skills, such as negotiating the obstacle course, advanced obedience, and laser directionals.

To train efficiently and effectively, we should make a list of all the behaviours we want the dog to be able to perform, and then make a schedule, which allows us to train each of these behaviours once a week, which may be over a longer or a shorter period.

We cannot attempt to do everything on the same day, but through the week, or even scheduling over a month, we can make a plan that we will train each skill once or more often depending on its priority, and how difficult it is. Once a new behaviour is learnt, it needs maintenance training once a month to keep it active.

I would like to see further studies on the frequency and duration of training, but for now we have some interesting leads.

ATTAINING EXCELLENCE

Even the best soccer players in the world train regularly to maintain and improve their skills. They know how to pass a ball, how to receive, shoot and defend, yet they practise every day because these fundamentals ensure their skills are readily available and fluent to the point where they appear effortless, and can be performed without conscious thought. It's like driving a car. In the beginning, it is difficult to operate the pedals and gears, adjust the speed, brake appropriately and use the steering wheel. But after a while all of this becomes second nature, and we are sometimes surprised when we reach our destination without realising how we got there. If a dog reaches this level of fluency in training, it is evidence of a strong mental bond between human and canine. The dog reacts quickly and reliably to the cues given; he can ignore distractions and is able to continue with his assigned task.

But no matter how well we train our dogs, they are not computers. When a cue is given, the output is not always what we intended, or trained, referred to as the black box (see *page 124*). When Thorndike was running experiments with his famous puzzle box, he showed us that animals change their behaviour because of its consequences. He called it the law of effect, but we should bear in mind that these experiments were conducted in a laboratory setting and the cat, who was involved in the experiment, was not facing everyday or operational distractions. It is the response to these 'happenings' that can affect outcomes in a way we might not expect.

Our dogs don't live in a Skinner laboratory; the world around them is full of distractions and their senses are bombarded with smells, sounds, sights and tastes. Dogs are always learning – 24 hours a day, seven days a week – and this will impact on their decision-making. A dog will be influenced by the reinforcers he receives, the presence of certain people or the environment he is working in. Whether the task is jumping a hurdle, storming through a tunnel, catching a frisbee, searching for narcotics, locating explosives, confronting a suspect or following a track, it starts in the dog's mind – and we have no idea of exactly what is going on in there. New technologies, such as MRI scans, reveal some the hidden mysteries – such as the intense brain activity recorded in detection dogs when they were anaethetised and presented with an imprinted target odour – but there is much to learn.

When we are training at an advanced level, it is important to be aware of the black box and to adapt training plans accordingly. This generally involves working on a lot of repetitions in order to establish a predictable outcome. For example, a radio directional dog that is doing 100 or more directional exercises may chase a rabbit when it suddenly appears in exercise number 86. We would, therefore, need to change our training plan and make sure that the dog is faced with rabbits in 86 of the 100 trials. The probability that the dog will be able to cope with a 'rabbit distraction' when he is on operational duty is now much higher. Imagine if the dog has seen rabbits in 1486 of the 1500 trials? We can be pretty confident he will not chase rabbits if he meets them when he is working.

KEEP-GOING SIGNAL (KGS)

When putting on shows at the *IQ Zoo*, Bob and Marian Bailey used 'keep going' signals to encourage animals to perform trained behaviours over an extended period of time. Music was used as an accompaniment, not simply for the entertainment of the spectators, but as a KGS for the animals. The animals knew that they had to work if the music was playing in order to get a bridge signal followed by a reinforcer. The importance and effectiveness of keep going signals was underlined in later work carried out for the US Navy in the 1960s. Dolphins were instructed to follow a specific route using a KGS. The training for this was simple and sophisticated at the same time:

- The dolphin heard a continuous keep going signal while it was swimming in the open sea.
- As soon as it left its course, the signal stopped.
- As soon the dolphin was back on course, the signal started again.

Using this technique, dolphins were steered to a destination – usually a ship – where they received a signal to carry out a specific task. For example, the dolphin might be fitted with a magnetic item and, on reaching the ship, it had to attach the item to the ship's hull. The initial training took place in small-scale areas with large targets. Observed by a trainer in a lookout tower, the dolphin was sent to the targets, following cues to change direction. When it reached the right target, it had to flip a lever. The trainer could see from above the water which target the dolphin had pressed and could reinforce accordingly. After many repetitions, and when strong conditioning had been achieved, the dolphins were moved to a larger training area and ultimately to the open sea. Even in their natural habitat, with its many distractions, the training was solid. The dolphins could complete 12-hour missions over 20 miles from their starting point, the mother ship – a level of training that still amazes me!

Keep-going signals are an effective tool when working with dogs. As an example:

- A continuous tone tells the dog to keep moving.
- As soon as he speeds up, the sound stops.
- The sound is only be switched on again if the dog is walking at the desired pace.
- Alternatively, the sound is stopped. The dog halts and looks to the left and right.
- As soon as he looks in the desired direction, the sound is switched on again and the dog continues on a new course.

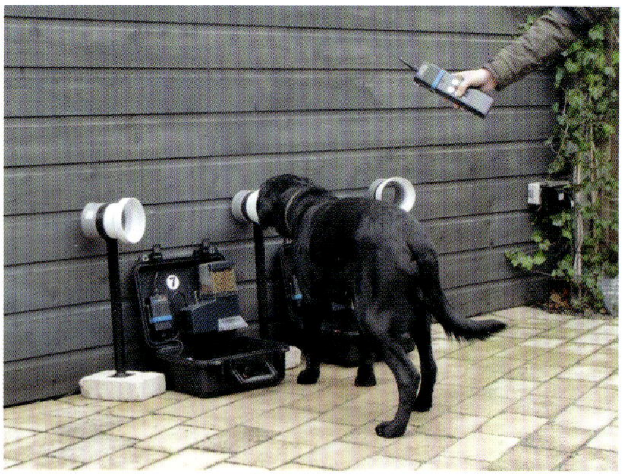

The set-up for basic training with the keep-going signal: The dog stays on the target for as long as he hears the sound signal.

Increasingly, the keep-going signal is used, more as an encouragement to the animal: *keep doing what you are doing to get the bridge signal and the reward*. Many trainers use some form of keep-going signal without realising it, especially when the dog is learning something new and the trainer wants to encourage him to maintain the behaviour. This might be:

- Keep holding the dummy.
- Keep walking next to the bike.
- Stay in a sit.
- Keep ignoring the distraction, etc.

I see the keep-going signal as a very powerful tool in training, particularly when working with issues such as distraction and duration, and I am surprised it is not put to more effective use. For example, let's imagine we want to practise a 'sit' over an extended period in a very distracting and challenging environment, such as in front of a busy supermarket. The dog sits but breaks the behaviour if the distractions are too great or the time span is too long. This issue can be resolved with a keep going signal. As soon as the dog sits, the KGS – a continuous tone, for example – starts. After a length of time, the KGS is followed by a bridge and reinforcement. When the dog breaks the behaviour, the keep-going signal stops and a reset will follow. Trainers often make use of a repeated and softly-spoken word, such as *"good"*, which has the same function as an audible signal.

SETTING BOUNDARIES

In our house, we have formulated some basic rules around feeding times. We use the time needed to prepare the food, and fill the bowls, to teach our two dogs to be calm and patient. We ask them to *"sit"* and put the bowls at a distance of 10m. We then call one of the dogs by name, which is their cue to run to their bowl. The dog then freezes and waits for further instructions. We repeat the same procedure with the other dog. The dogs must wait for our release cue, *"ok!"*, before they are allowed to pounce on their food and eat.

This behaviour did not come about because we would have punished the dogs if they had gone to their food without the release cue. Rather, it was achieved through setting rules, imposing boundaries and being 100 per cent clear in what we were asking for. Unfortunately, we humans often lack clarity when we are interacting with animals. Our emotions, our love, our frustration, our misunderstanding, our lack of communication, or our fear and passivity, send mixed signals. If, additionally, our body language is not congruent, we make it even harder. This is all exacerbated by our lack of consistency. For example, we tend to be more consistent when we are being watched by an instructor, but if our dog shows unwanted behaviour while we are chatting with friends, we don't take it too seriously.

How can a dog understand that:

- Pulling on the leash is suddenly undesirable when he has been allowed to do it for the last 12 months?
- Jumping up at people is ok, but not when they are wearing their best clothes?

If we want to turn unwanted behaviour into wanted behaviour, we need to train it and then be consistent. The first thing we need to do is establish clear rules and boundaries, which is easier said than done. The best plan is to write down the rules and then give them to someone else to read. Do the rules make sense? Are they easy to understand? If not, we need to keep working on them. Only when they are completely clear can we incorporate them into a training programme. When doing this, we need to bear in mind the *'what's in it for me'* principle (*see page 49*) to ensure the dog will find it worthwhile to co-operate. In this instance, it is certainly useful to have a dog who is obsessed by food!

Clarity, communication, and consistency is key. We not only need to use the same commands/cues, we also need to make sure that our body language is always the same. We must, at all costs, avoid sending mixed signals. There must also be respect. This is something that cannot be forced into a relationship – it must be earnt. We do not need to threaten or, worse still, to be physically violent (which tends to destroy respect). All we need to do is set rules and insist on compliance.

Of course, when we started training our dogs to wait for their food, they made mistakes:

- They got up from the 'sit' because they were impatient for the food.
- Sometimes they were reluctant to move when their name was called.
- There were also moments when they didn't wait to be called and crept towards their bowls.
- The freeze in front of the food bowl was a challenge – as was waiting for the release cue.

During the training process, we never resorted to punishment. We simply reset, going back to the point where the dog had been successful, and trying again. This brought total clarity to the training process. Remember, if we work with clear rules from the very beginning, it will be much easier – if not easy – to achieve our goal.

Rules and boundaries give our dogs a sense of security and wellbeing; they play a crucial role in animal welfare. Problems arise, and get worse, when they go unchecked. Imagine what it would be like to walk a dog who

- Attacks other dogs.
- Hunts wild animals in the woods.
- Hates cats.

Rules, boundaries and communication have to be trained.

- Continually scavenges (which can be dangerous for him).
- Barks incessantly.
- Fails to master basic obedience (which can be a danger for both road users and the dog himself).

It's sad, but many dogs spend extended periods in their crates – separated from their families – because they have not been taught the rules. If guardians fail to set boundaries, and unwanted behaviour has been allowed to escalate, the dog may be rehomed or, worst case scenario, be euthanised. Rules and boundaries do not inhibit freedom – they allow dogs to live their best life.

Consider what happens when no boundaries are in place:

- The dog who scavenges may need surgery because he has eaten something harmful.
- The dog who chases a cat and fails to respond to his recall cue, may be killed by a car.

In contrast, the dog who responds to basic rules and understands boundaries can be given the freedom to run free and have fun.

However, rules should not be seen as one size fits all. Setting limits and defining clear rules depends on the individuals involved, and it can be challenging to get it right. This is where professional trainers can help: they are experts in behaviour analysis, they can see the problem(s) and give guidance on how to proceed. Remember, a quick fix is not always the best medicine for unwanted behaviour.

For more information on setting boundaries, see page 157.

STIMULUS DISCRIMINATION (SD)

If we understand that stimuli can evoke a response – a behaviour – then we can imagine that they are key in giving us control over a behaviour. Stimuli are the triggers that spark behaviour. Of course, the animal must be able to perceive these stimuli. A dog cannot see infra-red light like some snakes, or hear low-frequency sounds like elephants, but there are plenty of stimuli he can perceive. But how can we train a dog to respond to a specific stimulus? For example, we want the dog to:

- Indicate a target odour when he is in an environment imbued with scent.
- Listen to his trainer/guardian's voice when he is in an environment full of human voices.
- Bite the suspect and not bystanders who happen to be present.
- Find an object that smells like a specific person.

All of this is only possible with the use of stimulus discrimination (SD). The procedure is simple:

- If the SD is given and a certain behaviour follows, it is reinforced.
- If the animal shows the same behaviour in response to a different stimulus, i.e., the SD is not present, it will not be reinforced.

Let's consider the example of teaching a dog to sit:

- The signal to sit is raising the right hand, with the palm facing the dog.
- This is repeated and reinforced several times.
- Now the trainer raises their left hand, with the palm facing the dog.
- If the dog sits, he is not reinforced. He is only reinforced when the signal is given with the right hand.

The dog has to alert on the target scent and to ignore other odours in the line-up.

The same procedure is used in scent detection training:

- In a line-up with eight stations, only one is loaded with the target scent. The rest are empty.
- If the dog indicates on that one scent, he will be reinforced.
- This is repeated and reinforced several times, with the added challenge of switching the target scent to different stations in the line-up.
- A second odour is added to the line-up, which now consists of the target scent, a new odour, and six empty stations.
- Of course, the dog will examine the new odour, but he will not be reinforced.
- Once he moves on and finds the target scent, he will be reinforced.
- Following several repetitions where the dog is reinforced for making the correct choice, each station, in turn, is loaded with a new odour, and just one station is loaded with the target scent.
- The dog understands that only the target scent (SD) leads to a reward. He will, therefore, learn to ignore all the other odours (stimuli) because they do not lead to a reward.

TIME OUT

This training technique involves a break in training and is usually employed when a dog is struggling with an exercise. There are many differences in how a time out is used. It could involve:

- Putting the dog in his crate.
- Suspending training for 15 minutes, or longer, before continuing.
- Putting the dog in a 'down' for five minutes
- Covering the dog's eyes before continuing.

The purpose of the time out is to get rid of unwanted behaviour. If the same time out routine has been used several times in a row, and the dog's behaviour has not changed, then it is not the right thing to do. To be effective, a time out must involve taking away positive reinforcement for the dog. Here we can make a comparison with raising children. A child who misbehaves may be sent to their room for a while, but this will have no effect on the unwanted behaviour if the room is full of positive reinforcements, such a toys, a tv, a play-station, etc. If a dog is taken to his crate and he finds a chew, water, a comfortable pillow, and his toys in there, he us unlikely to change the unwanted behaviour he has been exhibiting elsewhere.

A time out means 'no reinforcer' and begins when an unwanted behaviour is exhibited. It may only last a moment - maybe a few minutes - but it must result in a change in behaviour. It is not always necessary to take the dog somewhere else; a time out can work equally well if it takes place in the location where the unwanted behaviour occurred. For example, the dog can be put in a 'down' and simply ignored for a few seconds. However, this will only be effective where there is a strong bond between dog and trainer/guardian, and if the reinforcer (currently withdrawn) is especially motivating, such as food, toys, or attention (social reinforcer).

I use a time out in the form of a reset. I stop the dog when he is showing an unwanted behaviour and bring him back under my control, using either a verbal cue, or a long leash, and go back to my starting position. In other words, I close the window of opportunity for getting reinforcement. I then take a break of 5-10 seconds before starting again. Now the dog is strongly motivated to work for me, and to get the reinforcement that is important to him.

A time out can also be a punishment. If the time out procedure means putting a dog in his crate, he may end up disliking his crate in the future. For this reason, I use a platform/placement board when I am training, which is a most versatile piece of equipment. Primarily, I use it as a station; the dog learns that he must stay on the platform until I release him to do something else. Therefore, the platform can be seen as the starting point for a new behaviour. The length of time the dog remains on the platform is variable - it could be for as long as an hour if I am training another dog or processing data. After a number of training sessions, the platform starts to become a mental anchor for the dog; he will find rest, both physically and mentally. Because the platform is elevated - just above ground height - it provides clear criteria for the dog to stay in position. This, along with positive reinforcement, such as food or play, and because it is the starting point for a new exercise, makes the platform a highly desirable place to be.

Now the dog has a positive association with the platform, we can use it for a time out. If something goes wrong in training, I place the dog on the platform for just a few seconds - a time out - and restart the exercise. This is what I call a 'reset'. For example, the dog starts a line-up exercise from the platform. He makes a false positive on one of the distraction odours, so I stop the trial and place the dog on the platform as a time out/reset. After 5-10 seconds, I restart the dog.

In my experience, the time out is a useful safe and effective tool for changing behaviour, so long as there is clarity as to what is reinforcing for the dog in question.

THE MATCHING LAW

The law of adjustment or agreement

The matching law was formulated in 1961 by American psychologist, R. J. Herrnstein, and holds us accountable for the training we carry out. Data also plays an important role. To understand the matching law, let's consider the following training scenario.

If we practise 'sit' more often than 'down', and 'here' even less often, there is a good chance that our dog will perform a 'sit' better, faster, and more consistently than 'down' and 'here'. The same applies to dual purpose police dogs; training needs to be kept in balance if the dog is to be equally good at bite work and scent detection. It is important to bear in mind that this law is not only about the amount of training spent training 'sit', 'down' or 'here', it also involves the dog's energy and motivation for the task. If he responds quickly to the 'sit' cue – fulfilling all the necessary criteria – and not so well to the other obedience commands, we know that, going forward, we need to spend less time practising 'sit' and more time working on the other commands. Equally, if a police dog loves guard dog work but is less fond of searching for explosives, we know that we need to do more search work and only occasionally focus on bite work training.

The matching law also kicks in when there is a difference in the rate of reinforcement. If the reinforcement rate for 'sit' training is higher than for 'down' or 'here', the 'sit' will get a better response. The value of the reinforcer also plays a role. If the 'sit' is rewarded with plain kibble and the 'down' with fresh meat, alarm bells should start ringing!

The matching law shows us how a dog evaluates a situation. When implemented with thought and planning, evaluating and recording its impact, it is beneficial in training. Bear in mind, a dog will constantly adjust the behaviour that we request based on how well he was reinforced for it in the past. This means that a reinforcement history can work for and against us, as highlighted in the pros and cons of shaping and luring (see *page* 97).

It works in our favour in a familiar training environment as the dog knows this is where reinforcement happens. He recognises the training ground by scent, sound, and sight, and he gets a boost in energy, anticipating positive reinforcements that lie ahead. For him, the game has begun!

It can also work against us, as previously discussed in relation to training scent detection dogs. These dogs receive a high rate of reinforcement in the training environment, with lots of 'finds', but this changes when they are working on operations where they may have to work for long periods without finding the target scent and getting reinforcement – and sometimes there is nothing to find. This can be resolved by using the matching law principle and designing training in the way that matches operational scenarios. We need to strike a balance and set expectations – and yes, that does means longer search times and less finds when the dogs become more experienced and are ready for operational duty.

A dog will constantly adjust his behaviour based on how well he has been reinforced in the past.

DESENSITISATION

Just like us, dogs experience extreme emotions which can prove to be challenging in everyday life. There are dogs who are fearful. Sometimes the reason is obvious, such as a loud, unexpected noise, but it can also be an odd-looking person, a slippery floor, or an unfamiliar object on the path that causes the dog to freeze, bark, or even run away. On the other end of the spectrum, there are dogs who develop obsessions that override all other considerations. This could take the form of chasing cars or bikes, nipping joggers or herding small children. In some cases, the obsession is found in the home, such as the sound and movement of the vacuum cleaner or the doorbell. There are also dogs who have such a strong prey drive they chase every animal they come across, putting themselves and others at risk.

Desensitisation is a training process that can produce a different state of mind - but it is one of the most difficult techniques to master. To be effective, it needs close observation on the part of the trainer. It is a process that cannot be rushed and, often, it is the trainer's frustration that damages progression. Planning is, therefore, key to success. This is highlighted in the following example:

I was working with a camera dog who had a fear of slippery floors. Most of the time I simply avoided the problem, asking a colleague to deploy his dog in situations where slippery floors were involved. However, I wanted to find a solution.

The dog was more play-driven then food driven, but as I thought through the issue, I decided that attempting to instigate play on slippery floors was unlikely to lead to a positive outcome. I risked harming our relationship and the value of the play reinforcers. I also discarded the idea of breaking up the slippery floor surface with other materials, as I knew the dog would start to look for the 'safe' spots when he was working in buildings. After much thought, planning and preparation, I opted to try desensitisation.

My plan was to dip the dog's paws in a substance that would give him more grip on slippery floors. It would cost me both time and money, but the real issue was not allowing myself to become frustrated. In quick, dynamic deployments, time is of the essence, and I knew that proceeding slowly, to gradually desensitise the dog, would be a real challenge. Stopping a whole team to prepare the dog's paws was something I did not want to do. But, in fact, it turned out perfectly. Once I applied the substance, the dog had grip; he felt comfortable and was able to focus and work effectively inside buildings with slippery floors. The team had no problem with waiting a few seconds because the dog's role was invaluable. In time, I no longer needed to use the substance on his paws, and he never showed fear of slippery floors again.

Desensitisation has been used with considerable success when working with zoo animals. Historically, routine care and medical procedures could only take place if the animal was squeezed into a box or drugged with an anaesthetic gun. Now, with the help of desensitisation, a skilled trainer can train an animal to voluntarily participate in routine care, such as teeth cleaning and claw clipping, as well as tolerating medical interventions, such as accepting an injection with a syringe needle. In some cases, tolerating these procedures can make the difference between life and death.

Desensitisation must be undertaken in small, incremental stages. For example, when working towards administering an injection the trainer would teach the animal to:

- Come to the fence or cage rails at a signal.
- Spend extended periods of time in this location.
- Accept being touched with a plastic object on one area of its body.
- Stay in position as the object gradually becomes smaller and sharper.
- Finally accept the needle penetrating its skin.

Each stage - no matter how small - is rewarded and reinforced. The last stage, when the injection is administered, is accepted because the animal anticipates the reward it will be given.

In general, desensitisation is useful when a strong stimulus is present that either repels the animal or attracts it too strongly. The process involves lowering the value of the stimulus by first presenting it at a very low intensity and progressing in very small steps.

COUNTER CONDITIONING

Dogs don't choose to be scared. It just happens. Fear, along with other emotions, is hard to control and even harder to change. The philosopher, Spinoza, wrote: *"An emotion can only be controlled or destroyed by another emotion contrary to it."*

In other words, the only thing that can neutralise an impulse is a contrary impulse. The basis of counter conditioning is to work with opposing stimuli, getting rid of unwanted behaviour by turning it into behaviour that is acceptable/desirable. Food or play are chiefly used for this purpose. For example, a dog who is afraid of slippery floors learns to enjoy walking over them when he is motivated to play or given particularly tasty food. To explain how this achieved, let's consider a dog who is fearful of meeting other dogs when he is out on a walk:

- The conditioned stimulus (CS) is another dog.
- The conditioned (unwanted) response (CR) is fear.
- The unconditioned stimulus (US) is food.
- The unconditioned (desirable) response (UR) is a change to positive behaviour

If we want to get rid of the fear response using counter conditioning, we would:

- Combine the sight of another dog (CS) with the unconditioned stimulus (US) - tasty treats.
- This brings about a change in the dog's feelings and, if this unconditioned response (UR) is strong enough, he will no longer experience fear (CR) when he sees another dog.
- Instead, he will make a new connection between the other dog (CS) and the tasty treats (US) which will make him feel good.

It is important that the dog does not switch his focus to his guardian when he sees another dog. This would merely be a management strategy to cope with the encounter; it would not change the dog's attitude towards other dogs. The goal is to change the dog's state of mind from negative to positive, and this can be achieved by counter conditioning.

COUNTER CONDITIONING AND DESENSITISATION

When working with opposing stimuli, we make use of a counter measure to reduce the value of a stimulus, which involves a process of desensitisation. The difference between desensitisation and counter conditioning can be quickly determined by observing the trainer. When using counter conditioning, the trainer plays a much more active role, whereas desensitisation involves presenting the stimulus at a low intensity and gradually increasing it. Counter conditioning demands a high rate of reinforcement and perfect mechanical skills. Desensitisation is about patience, slowing down, going with the flow, and taking really small steps.

Of course, counter conditioning can also work the other way, changing a state of mind from positive to negative. In 1950, the psychologist, Dr. John Garcia, conducted a series of experiments with wolves to see if he could stop them killing sheep. He gave them poisoned sheep meat and they became ill. After they recovered, their behaviour changed - they stopped attacking sheep. The behaviour of the sheep also changed when they realised they were no longer at risk.

When working with counter conditioning, we make use of every opportunity to neutralise the high-value stimulus. Once a balance is reached, we can give more weight and value to the 'desirable' stimulus, i.e., the stimulus that is important to us. As a result, when both opposing stimuli are present, the dog will choose the one we like better. This practice is widespread in scent detection training when we want dogs to ignore distractions. For example, we task a dog to ignore food, but report narcotics. The stimulus - food while searching - must be of lesser value than the stimulus - scent of narcotics. For the trainer, this means making the scent of narcotics super interesting, so the dog ignores the food distraction.

A while ago we had a talented young scent detection dog in training. He was ten months old and displayed a tremendous prey drive which he directed at kongs. If he saw or smelled a kong, he would grab it and would be very reluctant to give it back. Instead of challenging this behaviour - trying to take the kong from him - we gave him *five* kongs which we left lying around in the training area. Obviously, there was no way he could put all five kongs in his mouth at the same time. As soon as he approached his handler, he was given a cue, followed by a great playtime with a kong on a rope. He quickly found out that the game only started on the handler's cue, and the kongs lying around were of little interest. The only kong he cared about was the one presented on the handler's cue. Now they could play together, and the dog was prepared to give up the kong on request because the value of the stimulus had changed. The game with his handler had now become more valuable; in effect, we had changed the power of the stimulus. We started with a situation where the kong was so high in value the dog would do anything to get it and had no intention of giving it up. To change his behaviour we had to find something that was even more valuable - and this was the interactive game with his handler.

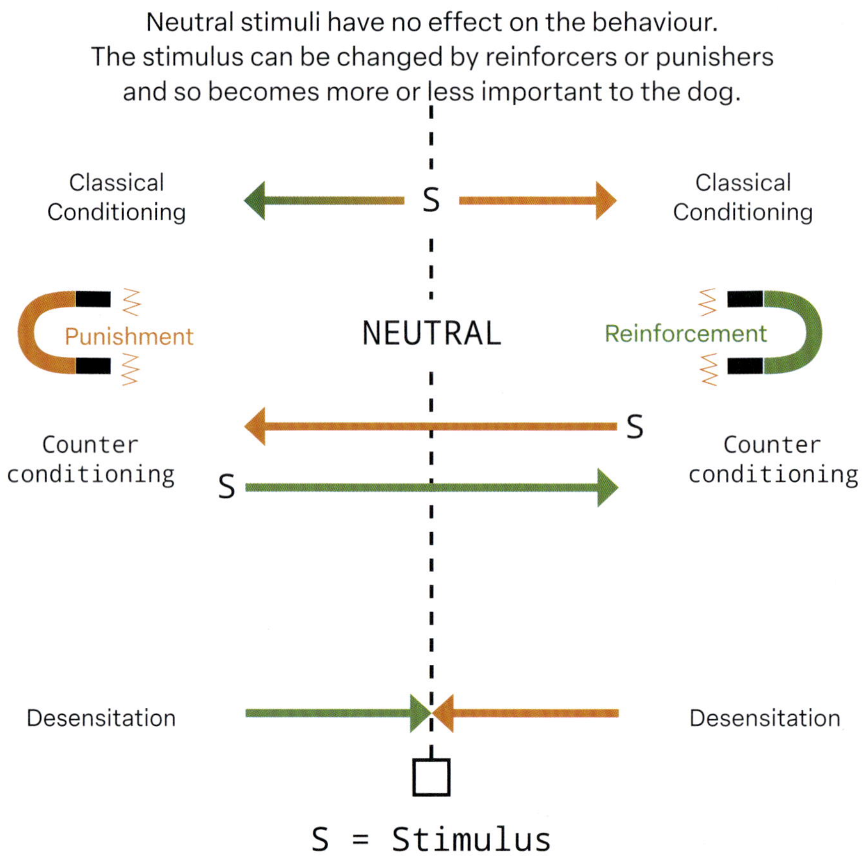

How the meaning of stimuli can change.

The same process can be used in relation to a dog who is frightened of loud noises, such as gunfire. We can gradually introduce the sound of gunfire from a distance, keeping the dog busy with food or with play. He needs to be focused on these good things so he is not paying attention to the gunfire. In this situation, we need to manage the situation so that the gunfire gets closer and louder, reverts to being at more of a distance and fainter, and then allow it to come closer and louder again. The aim is for the dog to connect the sound of gunfire with good things.

I found this process particularly useful in the early stages of training dogs to work in conditions where real explosives were used. It was not only the loud noise that was an issue, the dogs were also affected by the shock waves and the loose material that flew out when doors or windows were blown up. We, therefore, started with a small explosion and added some wonderful reinforcing events after it. In this way, the explosion became a trigger, as the dogs anticipated the good things to come. I remember an exercise where a door was blown up using too many explosives.

Traveling by helicopter needs step-by-step training.

We were somewhat overwhelmed by the enormous explosion – but my camera dog was already running at full speed towards the building! With repeated exposure, the dogs learn that a loud bang means action – and they can hardly wait to make their way through the smoldering opening where the door or window has been blown out.

I also used counter conditioning when we were training dogs to travel in helicopters. Initially we had problems with dogs boarding stand-by helicopters as they were not only worried by the sound of the engine, but also by spinning rotor blades, the smell of gasoline and the vibration. We adopted the following strategy, working with the dogs on an individual basis:

- We started by approaching a stationary helicopter, with the engine switched off, and engaging the dog in a game.
- After a few rehearsals, we boarded the helicopter and the pilot switched on the engine – but no spinning rotor blades at this stage.
- We stayed on board for a few minutes and then stepped off and played a game.
- Step in and out, play a game, on repeat.
- Then, applying the same rules, the rotor blades were activated, followed by a take-off and then a short flight, ensuring that each step was reinforced with play.

The result of progressing in small steps, and creating the positive association with play, was that the dogs soon learnt to love their helicopter rides.

FLOODING

Flooding is an extreme form of desensitisation. It's fast and can be very effective but, as already highlighted, if used incorrectly, it has the potential of being very dangerous (see *page 132).* The process involves exposing the animal to the stimulus he fears, and keeping him in that situation, position, or environment until he 'gives up', and the stimulus is no longer a source of fear. However, there is always a risk that the animal becomes even more fearful and, as a result, slips into a state of learned helplessness (see *page 135).*

I have used flooding as an effective training technique when introducing dogs to abseiling. The dog is fitted with a harness and is then roped together with his handler. The handler steps down from the practice platform, or the edge of the building, taking the dog with them. Most dogs experience a brief moment of panic, kicking or attempting to get back on to the platform. But after a few seconds, the dog will calm down and accept the situation. After some praise and petting from the handler, the pair start abseiling – and most dogs love it.

The benefit of flooding is that if a dog overcomes his fear using this technique, he will not experience that fear in similar situations. For example, flooding can be used when accustoming dogs to confined spaces, such as a small box, a narrow, noisy motorboat, or similar uncomfortable, noisy spaces. The dog will resist at first, but then quickly calm down and accept the new situation. I have used flooding in some very challenging situations, such as climbing very difficult stairs, entering ship engine rooms or in work situations with a large number of people and animals on the move.

Once I was working on the roof of a high building, which was reached by a flight of narrow metal stairs. My dog, Shadow, was well trained with stairs, but he had an issue with this staircase, and he was in danger of jeopardising the operation. I waited until I was in a safe zone and used a simple flooding technique on a similar flight of stairs. As soon as Shadow realised he was able to climb the stairs, he lost his fear. We were able to continue with the operation, as planned, and Shadow showed no concern when he was asked to go up and down the metal stairs.

When used by experienced trainers, flooding can be very helpful and, in some extreme cases, it can even be life-saving. Dogs have been euthanasied because their fear has been so debilitating, but this may not have been necessary if flooding had been introduced by an expert. The greatest challenge in relation to flooding is that once it has been started, it cannot be stopped halfway through. The stimulus cannot be removed until the dog is calm and desensitised. It is also important to be aware that a dog may react aggressively in this phase and the relationship between dog and trainer could be damaged. Flooding is not for beginners but, in the right hands, it is an effective training tool.

REDIRECT TO A DIFFERENT BEHAVIOUR

Instead of punishing a dog for unwanted behaviour, he can be redirected to other behaviours that can be reinforced. As a result, the dog chooses the behaviour that leads to reinforcement and is no longer concerned with the unwanted behaviour. The key is to ensure that the unwanted behaviour cannot be performed at the same time as the replacement behaviour. Therefore, the alternative behaviour, chosen by the trainer, must be incompatible with the undesirable behaviour. For example:

- A dog who jumps up on visitors will be sent to his bed when the doorbell rings and will receive a great reward for staying there.
- A dog who begs at the table will be sent to his crate during mealtimes and will find a tasty food reward waiting for him.
- When we trained dogs to crawl with us in a prone/crouching position, we used a hand-held target held close to the ground – a white sticker on a black glove – and reinforced them. This meant the dogs were busy and would not jump up and run to where the action was.

There are instances when we do not need a to provide a specific, alternative behaviour that is incompatible with the initial, unwanted behaviour.

If a dog is very focused on a certain behaviour, such as watching out for squirrels, we could reinforce him for looking at anything other than squirrels. This is best done by throwing food – tossing treats the moment the dog switches focus – glancing up at his guardian, sighting a dog in the distance, sniffing, sitting or lying down – the behaviour itself is immaterial, as long as he is no longer looking out for squirrels..

FEAR AND ANXIETY

In order to learn and progress, a dog not only needs to acquire the necessary skills, he also needs to be in the right frame of mind. If a dog suffers from recurring anxiety, training sessions can be a real trial, even if he is capable of carrying out the exercises he is asked to do. Fear and anxiety present major challenges for both dog and handler and, in most cases, professional help is required. A combination of exercise, medication, and a change in environment or daily routine can often provide relief. However, we need to be aware that although management and medication can help, it does not address the cause.

Some of these problems are the result of poor breeding practices, such as a failure to prioritise sound temperament in breeding stock, or lack of socialisation when puppies are being reared. When puppies are raised by unscrupulous breeders in puppy farms and see nothing but their mother and siblings for twelve weeks, it will have dire consequences. Bad handling can also play a part. Puppies are not born with an innate fear of vacuums, vets, or young children. Problems that arise are the result of classical conditioning, when puppies have learnt that these things may cause pain or discomfort:

- When that nice man in the white coat suddenly sticks a hypodermic needle in your bottom, it hurts.
- When a vacuum cleaner comes too close when you are sleeping, it's frightening.
- When a toddler is chasing you, there is good reason to be scared.

To resolve these and similar issues, we need to give guidance by doing nothing. If a guardian pays attention to the concern – worrying because the dog will experience pain when he is injected – it is likely to become a much bigger problem. So, yes, a needle will hurt, but we don't focus on it. Prior to the event, we distract the dog with food or play, and continue giving tasty treats or playing immediately after the injection has been given.

We can adopt the same approach with a dog who is scared of the vacuum cleaner. All we need to do is switch it off, leave it on the floor and ignore it. Then play with dog or give him food in the vicinity of the vacuum cleaner. Now treat the dog while pretending to vacuum, i.e., with the vacuum cleaner switched off. Finally, switch on the vacuum cleaner and intersperse vacuuming and treating the dog. In this way, he will not only become accustomed to the vacuum cleaner, he will also start to associate it with good things.

In the case of a dog being chased by a toddler, this must be stopped immediately. No dog should be expected to tolerate this behaviour. The toddler should be placed in a high-chair, for example, while the dog is given treats or enjoys a an interactive game, paying no attention to the child. This can be progressed in small steps, until the dog is comfortable in the presence of the toddler.

The message is: don't make the needle, the vacuum cleaner or the toddler important. Make the dog important by playing with him or feeding tasty treats. This form of distraction, coupled with the guidance of assuming an everyday, commonsense manner, means the 'scary' event/situation loses its impact.

THE FASCINATING WORLD OF SCENT RECOGNITION

Scent discrimination is one of the most fascinating disciplines because we have to lay aside our desire to control a situation and rely entirely on the dog's nose.

Everything smells - almost every item has a unique olfactory picture. We humans are aware of a range of smells, but the dog's sense of smell is so vastly superior, he can detect odours that we are incapable of perceiving. A dog can not only sniff out explosives and narcotics, he can also detect the odour of money and electronic devices such as telephones, SIM cards, data sticks or hard drives - items that may be of particular interest to law enforcement personnel.

The dog's amazing sense of smell is a blessing and a curse, as he can detect a vast array of scents that interest him, which can be a huge distraction when tasked with finding a specific, target scent. For this reason, scent detection training has to be highly structured and disciplined. For example, if we are careless in hiding a target scent and contaminate it with our human scent, before we know it, we will have a dog who will only respond to items that have human scent attached to them. Now we could be smart and say: *"Ok, I'll only touch the items with disposable gloves"*, but watch out for pitfalls! It may be that the dog now understands that 'disposable latex gloves' is his target scent. So, working 'clean' means ensuring that the dog is only looking for the specific target scent that we have set before the start of the training - and that can be far from easy.

Consider what might be involved when a dog is tasked with finding and indicating on banknotes. What do we identify as the unique, distinctive target scent? Is it the special paper that banknotes are printed on, or maybe the special ink? Can a dog be trained to identify the scent relating exclusively to the printed security features, or do we need to train a combination of all these odours? There are many questions that need to be answered before we start training, because we cannot push a 'reset button' in the dog's brain once he has memorised a specific target scent.

So, it all starts with identifying the exact target scent we are going to work with. We want the dog to target the items that are of interest to us and avoid false alerts as far as possible. However, we are not always able to work with original substances - a civilian dog handler does not have access to cocaine, explosives or the ink used for printing banknotes. For this reason, substitute odours are manufactured for training purposes. Unfortunately, this is not as straightforward as it sounds. A dog trained on a substitute scent may not indicate the real substance if he encounters it in an operational situation. Imagine being offered two milkshakes in identical glasses. One is freshly mixed from real milk, real ice cream and bananas, the other comes from a fast-food restaurant. Of course, you can taste the difference!

All scents are made up of specific elements such as high or low evaporation pressure. We humans, with our vastly inferior sense of smell, can detect high evaporation scents, such as coffee or tobacco. In some cases, we can even work out where the scent is coming from – the scent of coffee wafting from the kitchen or cigarette smoke coming from a neighbouring balcony. Odours with low evaporation pressure, such as salt or sugar, are more difficult for us to perceive; we have to get our nose very close to notice the difference between them. Smells also arouse emotions in us because we associate memories with them – the perfume of an ex-lover, or the unmistakable smell encountered when entering a dental practice. Emotions linked to smells can be good or bad. When people have suffered trauma, a particular odour, such as the smell of burning, can trigger an extreme reaction.

Dogs are known for their superior olfactory system, which gives them a heightened sensitivity to smell. In scent detection training, we, therefore, try to avoid problems caused by using odours with an evaporation density that is too high or too low, or those that may have unpleasant associations. There are multiple 'starter' odours, such as a piece of a tennis ball, a few grams of vanilla, or a few drops of essential oils, that are both tried and tested. Many trainers generally favour using the scent of a kong, as it has positive expectations anchored in the dog's brain. In addition, it also has a very specific smell, especially when it is new. Even we humans can perceive it. The bonus is that it appears to be a unique rubber compound found nowhere else. When our dogs have been set to work in all sorts of training scenarios and operations, even in warehouses housing different types of rubber, or garages specialising in tyre repairs, they have still been able to find and identify small pieces of kong and did not gave false indications on any other rubber material.

Another great benefit of using the kong as a base target scent is that it can be transported anywhere in the world, which means I can take them on planes when I am traveling to training events. In fact, they are readily available in most countries which means I don't have to panic if I lose one. In addition, I can work with kongs in hot environments, which is a real bonus. Sunlight with UV has a big impact on odour pictures, changing them very quickly. Volatile molecules leave the odour source, and evaporate swiftly, so it is hard to detect the odour. A kong seems to have such a specific odour picture, which allows dogs to locate it in cold, normal and very hot conditions.

The material is non-toxic if chewed or even swallowed, and it can also be cut into very small pieces to increase the degree of difficulty for the dog.

The principle of training a target scent is that the smell of the kong (in this instance) is linked to pleasant emotions via classic conditioning (see *page 53*). We say that we 'imprint the dog on the smell'. In practice, we present him with target scent – a small piece of kong – in a cup, for example, and then click and reward as he smells it. Following a series of repetitions, the dog will be 'locked' on the kong smell, i.e., he finds it meaningful and pleasant. We can then set up a simple search. If the dog is successful, we can progress as follows:

We can practise getting finer and weaker odour concentrations by cutting the kong into smaller and smaller pieces.

Or, more difficult, we can transfer the

kong smell to another material, such as an odour carrier, and ask the dog to search for it.

Alternatively, we could use imprinting on the kong smell to train another, new target smell. Let's take a SIM card from a mobile phone as an example. This could be done with the help of a line-up:

- Initially, we present the dog with the known target smell - the kong (cut into small bits) in the scent container - and reinforce him.
- Once the dog has learnt that the kong smell gets him the reinforcement, we add the new smell (the SIM card) to the container with the bits of kong.
- Repeat and reinforce it a few times.
- We now take away the bits of kong, leaving the SIM card in the scent container. Now we need to be really quick with our bridge signal and our reward, because the dog may pause very briefly over the container and not show a complete alert. We need to reward the slightest hint of an indication!
- At this stage, the scent container still has traces of the kong smell from the cut-up bits that were in there previously. Therefore, the next step is to work with a new, clean odour container and a new SIM card (because the one used so far also smells like kong). Now the SIM card is the only target scent and becomes our 'hot target'.

In this exercise, we are helped by the fact that, as a result of classical conditioning, the dog perceives the kong smell as positive and transfers this positive association to the new smell - the SIM card. But how little scent does a dog need to make a positive identification? Here is the catch. The smell of kong is strong and

FIND THE KONG

Here Charlie is looking for a metal disc that smells minimally of kong (imperceptible to our noses) because it was stored in a jar with small pieces of kong. No problem for the dog's nose!

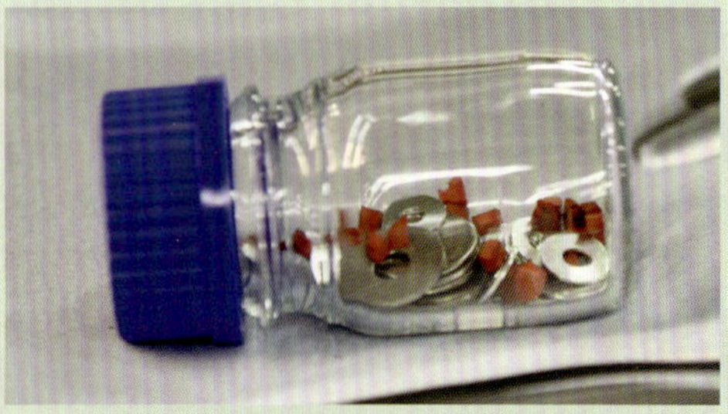

A metal disc is kept in a screw-top jar together with small pieces of a kong.

The disc prepared in this way, has acquired the kong smell, and is hidden behind the ashtray of a car...

... Charlie has no problems in finding it!

distinct; it is easily detected by a trained dog from a distance of over 150m in an open field. So how small do the pieces of kong need to be to train dogs to find the lowest thresholds of target odour? And how can we monitor their performances if we don't know what we actually are doing?

As well as trying to work out the science behind minimal thresholds, there is a practical consideration. Training a scent detection dog using the 'feel good' odour of a kong or a tennis ball is highly effective, but there is a drawback. Consider a dog searching through luggage in an airport. There is a reasonable chance that he may find a tennis ball, packed ready for holiday activities or, in my case, my suitcase may well contain multiple kongs if I am travelling to a workshop. So yes, there is a risk that a flight will be delayed because an explosive detection dog, previously trained on these odours, is alerting on a suitcase with a kong or tennis ball inside.

In addition, there is a difference between Europe and USA in terms of legal procedure. In Europe, if a detection dog gives an alert, further investigation is permitted and the court will not, at this stage, question the reliability of the training. However, in many instances in the USA, the K9 handler has to prove to the court that his dog is only trained on the target odour he is tasked to find in operations. This means explosive dogs can only be trained with explosives, and drug dogs can only be trained with narcotics, etc. The advantage is that operational alerts can never be questioned. In contrast, it could happen that detection dogs in Europe could react to a kong or tennis ball during a vehicle search instead of alerting on explosives or narcotics. However, smart trainers understand that we cannot delete or erase those odours from the dog's mind if they have been used in training.

NOVEL ODOUR TRAINING AID (NOTA)

To overcome these issues, an exciting new technique for training detection dogs is now being implemented, which employs a novel odour training aid (NOTA). This entails working with an odour that does not exist in the everyday world. It cannot be found in toyshops, pet shops or anywhere else. The odour is specially manufactured for training detection dogs and is used exclusively for this purpose. At the start of training, this scent is used for imprinting, training and testing, allowing dogs and handlers to prepare for the real-life scenarios without using dangerous, toxic or illegal products. NOTA comes in seven concentrations and also with a control odour. It is the perfect way to train and proof detection dogs and will revolutionise this aspect of scentwork.

When a dog has successfully completed scent detection training on NOTA, the decision can be made as to which specific target odour(s) to use for imprinting, i.e., establishing a target odour in operational situations. The big advantage of this technique is that there are no grey areas when a dog alerts on an odour because the NOTA does not exist in the real world. It cannot be found incidentally in luggage or when searching the inside of a car, therefore the dog's indication will be 100 per cent reliable – and no court of law can question it. In addition, maintenance training with NOTA is a useful option if a dog runs into problems his working life.

PASSIVE OR ACTIVE ALERT?

The alert tells us that the dog has found his target scent, and he can do this actively or passively by using different types of behaviour:

Active: As the name suggests, this may involve barking, jumping up, pawing, digging, or even biting the item. This form of alert is often seen in professional scent detection dog work, although it has many disadvantages – we don't want a drug detection dog to bite cocaine or an explosives dog to paw at a landmine.

Passive: This is more subtle form of behaviour where the dog will simply lie down or sit when he identifies the target scent, or he may 'freeze' and stare at the item in question. It is generally used in semi-professional scent detection work and in competition, even though it requires more work at the start of training and the trainer needs more skill. For dogs that go into action, this type of alert is the best and safest for everyone involved – and not only from a safety point of view. Think about the damage a narcotics dog can cause if he sinks his teeth into the dashboard of an expensive car or a piece of antique furniture.

The world of scent detection is one of the most fascinating areas of dog training. It is of tremendous importance when used professionally, but it's value when used in sport, or simply for fun, should not be under-estimated. Following a scent comes naturally to all dogs; they relish the opportunity to use their amazing olfactory system – and being rewarded for something they love to do!

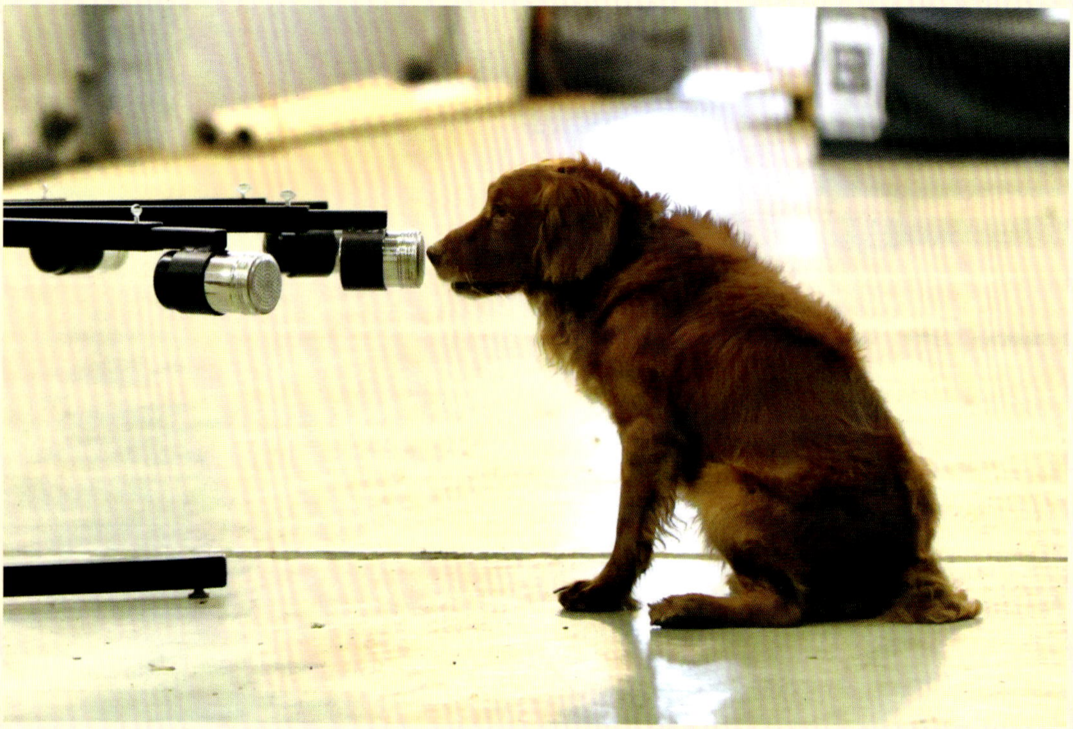

Passive alert: The dog sits or lies down in front of the target scent, looks at it and freezes.

Chapter 17:

WE NEED TO TALK ABOUT PUNISHMENT

When I am hosting a workshop or a seminar, I ask participants to raise their hands if they are using punishment in their training. Usually only one or two out of a hundred trainers present raise their hands. If I go on to ask who has worked with punishment in the past, maybe three or five will be added to the small minority of honest people. To this day, there is a burden of shame and guilt among trainers who have used punishment in the past and, this can be made worse by those who claim to use purely positive methods. However, as I have previously highlighted when discussing the four quadrants (see *Chapter Six*), aversives (*see Chapter Nine*), and working with positive and negative consequences (see *Chapter 12*) some form of punishment will always be part of the learning process:

- If a pupil is late for school for several days in a row, he will receive punishment in the form of a letter to his parents or a detention. A punitive technique is used to change his undesirable behaviour of being late.
- If a motorist is caught speeding, the result is a fine, or maybe losing their driver's licence, which acts as some form of deterrent.

We are familiar with both examples of punishment and are happy to discuss them openly. But what happens if:

- We lose money because we have failed to read the small print on a contract we have signed?
- We have to take a test to renew a driving licence which was lost after drink driving?

In these examples, we have been caught in a compromising situation, and we don't want to talk about it. We may feel guilty because we have done something bad, or we may feel shame because we think we are bad, and that makes us feel bad about ourselves. I fully admit that I have worked with punishment and coercion in the past. These techniques worked, but it did not feel right, so I started looking for different options. Today, my colleagues and I work with positive reinforcement, but we still use punishment as well. Yes, you read that right! Take a deep breath before you shut the book and toss it in the corner. Suspend the outrage and consider what we mean by punishment:

- You give your dog a cue, he doesn't meet the criteria and so you don't reinforce him – you punish him!
- Your dog has been behaving the way you want, but you have decided to go with variable reinforcement from now on – you punish him!

So if we accept that some form of punishment is part of the learning process, we need to consider:

- When to use punishment.
- How to use punishment.
- What are we trying to achieve?

Are there other ways to change behaviour? Maybe it's better to change the word 'punishment' to a term that is less emotive, such as 'correction' or 'blocking unwanted behaviour', so that we can have a rational discussion and learn from each other. Bear in mind, there is no difference between punishment, aversives and corrections. While 'correction' sounds more harmless than 'punishment', the goal is the same: to reduce or stop unwanted behaviour.

I have used corrections to change the behaviour of a dog that would otherwise have been euthanised as his guardians could no longer handle him. His unwanted behaviour - biting - had become so extreme that the only options were putting him to sleep or changing the unwanted behaviour through punishment (read correction). For this measure to be effective, it is essential to bear in mind that:

- Punishment is a last resort and should only considered when all other options have been exhausted.
- As previously outlined, there are different reinforcement schedules which promote the use of positive reinforcement (see *page 108*). If we find ourselves in a situation where we feel the need to use punishment, we should first review these schedules to see if there is another way forward.
- Punishment should only undertaken by an experienced, professional trainer.
- In the wrong hands, the use of punishment, can exacerbate behaviour, leading to issues such as aggression.
- If punishment is unavoidable, there is no room for frustration or anger on the part of the trainer.
- We should 'correct' only one specific behaviour. Remember, we are punishing the behaviour - not the dog.
- The correction should be carried out only once - the timing is vital.
- Punishment must be quick and effective, and of sufficient strength to stop the behaviour.
- Punishing 'a little', at a subliminal level, does not only ruin the relationship with the dog, it can at worst, make the unwanted behaviour even stronger by introducing an element of variable reinforcement.
- Consistency is key: rules do not apply on one day and not on the next - they apply every single day.

In the case relating to the dog with the biting issue, as soon as I stopped the undesirable behaviour, I started training an alternative behaviour. For example, the dog was sent to his crate before he got his food or before visitors entered the house. The dog learnt the new rules quickly; they gave him security and structure. Everyone could breathe a sigh of relief and the dog developed a very positive attitude. This meant his guardians had the opportunity to positively reinforce the 'new' behaviour, which was a major stepping-stone in improving their relationship with him.

Do we feel ashamed or guilty when we use punishment? Both feelings block an effective, open discussion. We are so preoccupied with debating the rights and wrongs, we forget what we are trying to achieve. If, on the other hand, we keep the discussion open and are willing to learn from each other, we can make the world a little better for trainers, guardians and dogs.

Do I like working with punishment? Certainly not! However, it can sometimes be a last resort in solving problems that people have created by not setting boundaries, or by unintentionally allowing undesirable behaviours to be repeated over and over again. Of course, before using

SHAME AND GUILT IN DOG TRAINERS

We live in a world where it is normal to judge others, or to be judged ourselves. We are bad at dealing with this because it puts us in a defensive position. Shame and guilt weigh us down and damage our self-esteem – a sensitive issue in the tough world of dog training. As soon as the discussion moves towards penalties, data collection or the effectiveness and efficiency of training, many coaches see it as a form of attack. They have learnt from their dogs that attack is the best form of defence.

It is imperative that trainers move forward in their understanding of punishment rather than feeling guilty about the past or pointing fingers at those who have yet to discover a more positive way of training. Research professor and podcast host, Brené Brown, has written a number of books on shame and guilt, which I have found most enlightening. Brené compares the experience of feeling shame and guilt to a swamp. The gremlins that live inside the swamp try to drag people downwards by telling by them that they have done poorly, made mistakes, and it's all their fault that things don't go as expected. So, in the context of animal training, let's emerge from the swamp of shame and guilt, and learn from each other.

punishment, we need to exhaust all other possibilities that training offers us. For example, if a dog chases cats, we can:

- Train a reliable recall.
- Work on focus with cats (at a safe distance).
- Work with a long leash.

However, if the dog has been successful (from his point of view) in grabbing and even killing a cat, he will not co-operate with trained behaviours in return for a piece of sausage, a ball game, or any other reinforcer (or removing a reinforcer) because the reinforcer – hunting success – is much more powerful.

Unwanted behaviour is a criteria or definition that we humans make, and not the dog. In other words, there are no ethical dogs. For most dogs, a steak on the kitchen table is a steak for them and they will try to grab it if they smell it. It the same for an attack dog; there is no difference between biting the jeans worn by a suspect, or the blue uniform trousers worn by a police officer who suddenly appears on the scene.

We also need to define what is a penalty and what is a reinforcer and, more importantly, we need to work out what the individual dog perceives as being reinforcing or punishing, and how this affects their behaviour. For example, a young Springer Spaniel I was training ran into a thicket of brambles in pursuit of a rabbit. I needed gloves and a hedge trimmer to free him, only to watch him jump right back in, desperate to find the rabbit. On a more extreme level, when I was working in Africa, I was tasked with putting a needle in the front paw of a lion that we had sedated for treatment.

No matter how hard I tried, I could not penetrate the paw because of the accumulation of scar tissue. Lions fight, get hurt, heal, and fight again because the reinforcer (food or a sexual partner) is so powerful it overrides the pain resulting from the injuries they suffer.

We share a belief that to gain something, we must pay for it, or suffer in some form. We cheerfully accept the premise of *no pain, no gain*. For example:

- Those who want to learn about dog training, will need to study some heavyweight books (including this one!).
- Those who want to gain more hands-on training experience, will need to spend time and energy working with their dog outside, perhaps in the rain and cold.
- Those who want to lose weight, will need to work out in the gym.

But whether reading a book, training a dog in the rain, or working out at the gym is punitive or rewarding depends on individual perceptions. Remember, when the behaviour increases, it's a reinforcer; when it decreases, it's a punishment.

When we step into territory where punishment/correction is being considered, we should not go further until we have worked through the *where, when, why and how*. We should ask ourselves the following questions:

- Are we employing effective, efficient, and ethical methods to achieve our goals?
- Are there other options?
- Have we sought expert advice?
- Do those experts have practical experience?
- Do their opinions align with our values and moral standards?
- Do our goals - and the steps to achieve them - comply with the findings of modern science and practical experience?
- Do we have confidence in our goals? Are they achievable?

MISTAKE OR UNWANTED BEHAVIOUR?

Within the training community, we need to have rational discussions as to what constitutes desirable and undesirable behaviour. Is a mistake an unwanted behaviour? And what is a mistake? If an explosives dog indicates a piece of baggage, the aircraft will be evacuated. But if, on further investigation, no explosives are found, is this a mistake on the dog's part, or unwanted behaviour? What about the police dog who bites a 10-year-old boy in error? Or the Rottweiler who digs his teeth in the postman's leg? And yes, I have also witnessed a trained assistance dog abandoning his guardian in a wheelchair to go off and hunt ducks in the city park.

We need to look at the behaviour itself:

- How do we classify it?
- What is the handler/trainer/guardian trying to achieve?
- What is acceptable and what is necessary?
- What will the end result look like?
- Are we happy with the outcome?

We live in a polarised world where everything is seen in black or white, left or right, positive reinforcement or coercion training. We need to make a choice; remaining neutral is unacceptable. Again, I would emphasise: I stay away from punishment if I can achieve my goals in other ways. I am an advocate for using positive reinforcement in training. However, a training process is not built purely on positive reinforcement; trial and error are essential elements of learning. The four quadrants in operant conditioning make this clear (see *page 59*). From a human perspective, there are those who will steal, drive after drinking, and do other bad things to other humans or animals. Sometimes it is necessary to change their unwanted behaviour by punishment in the form of a fine or even a spell in prison. In terms of learning by consequence, we are not so different from animals.

I believe there are only three reasons that justify using punishment. The unwanted behaviour will:

- Result in injury, or even death to the animal itself.
- Put colleagues, bystanders or other animals at risk.
- Pose a risk for the operation or the environment.

To put this in context: if a police dog is out of control and keeps biting the suspect, he will not only cause serious injury to the suspect, he may also suffer severe consequences if the suspect retaliates. The dog who is chasing rabbits may cause untold damage if he runs across a busy road and causes a pile-up.

Punishment is the last option I would choose, but if it's necessary to save the dog, other animals, or bystanders, it is an option that I will consider. We need to put emotions aside and look at behaviour and training from a scientific perspective. We can turn unwanted behaviour into wanted behaviour through training. Ignoring unwanted behaviour, or trying to avoid it, can make the problem worse. Sadly, this can result in shelters full of dogs who have been abandoned by their families, and for some, euthanasia is the only remaining option.

The examples of mistakes/unwanted behaviour, I outlined earlier, are from real life:

- The detection dog that alerted on the wrong bag cost the airline $250,000 and they terminated their contract with the K9 unit.
- In a police operation involving an armed hostage, a young boy got too close to a member of the task force and his patrol dog and was attacked.
- The Rottweiler, who had a history of barking from inside the front door, and was never taken seriously, took his opportunity when the door was left open and pounced on the postman.
- The disabled owner sought help from the organisation who trained his duck-chasing dog and was told to use high-value reinforcers us to change his behaviour rather than punishment. However, this made no difference and, tragically, the dog was drowned when he fell into deep water in pursuit of ducks.

Mistakes are made by both dogs and their human partners, and unwanted behaviours are exhibited by both. In the preceding scenarios, what could have been done differently?

- A second dog could have been used to confirm the alert but this also carries a risk. We know that dogs leave saliva and probably other odour cues at the place where they alert, so other dogs could use this as a cue to alert in the same location. In addition, what if there is a difference in threshold between the two dogs regarding their powers of detection? The first dog may be able detect very low quantities of odour which the second dog fails to identify.

- The police officer could have muzzled the dog.
- The Rottweiler could have been kept away from the front door.
- The assistance dog could have been placed with a client who lived in a city environment where he would probably never come across ducks and water.

Ultimately, it is a question of ethics as to whether we ignore, manage or solve problems of this type. It's all about integrity and quality of life.

Let us now consider the difference between a 'mistake' and 'unwanted behaviour' and work out whether they should have different outcomes.

A **mistake** is an action, decision, or judgment that leads to an unwanted or unintended result. While a mistake is unintentional, an error in judgement can be both with and without intent. Deliberate deception is, of course, something else.

Unwanted behaviour does not fit into the environment and cannot be accepted within the framework of norms, values and standards. For example, drinking and driving is an unwanted behaviour, not a mistake. The same applies to theft. If I steal a bike because I don't want to walk, it's not a mistake, it is unwanted behaviour!

The consequences that follow a mistake are influenced by the perpetrator's intention. The scent detection dog, who really believed that there were explosives in the luggage, could have made a mistake because a different odour picture was present which contained one of the components that are also found in explosives. For example, many explosives dogs in training indicate when they detect the smell of diesel because it is present in many homemade explosives. However, if the scent detection dog was simply looking for a reward and deliberately misled his handler with a false alert, the intention is different and, therefore, the consequences should also be different. But for a handler to make that call - deciding if was a genuine mistake, or unwanted behaviour - is very tough. The handler would need to have a detailed knowledge of the dog's behaviour and even so, there would still be an element of doubt. Ideally, we should collect and evaluate data to make a more objective decision (see *Chapter 20*).

To assess whether something is unwanted behaviour, we need to be clear about the norms and values that affect our decision-making. This assesment will be made by those responsible for the dog and his actions: the dog's guardian/handler/trainer, household members, the service unit, the sports club, etc., as they are in a position to decide what is acceptable or otherwise:

- It is desirable for a police dog to catch and apprehend an armed criminal, but not to attack a 10-year-old boy.
- Retrieving ducks is desirable behaviour for a Labrador Retriever - unless he has been trained as an assistance dog.

Unwanted behaviour is shaped by the norms and values of our human world, whereas a mistake is the result of a deliberate action/decision. In both cases it is difficult to choose the right consequence because it is not always clear what is going on and why. Not all stimuli that lead to a reaction are clearly recognisable to us humans. In such situations, guidance based on gut feeling is like trying to steer a car in thick fog. We need data!

If we look at the natural behaviour of dogs, we see that they can kill other animals to eat or fight with other dogs to defend their territory, but they can also be very gentle when caring for their puppies. In nature there is no discussion as to whether something is positive, balanced, or coercive. There are struggles for who will get the best mate, and punishment comes in the form of heat, cold, hunger, thirst, or attack. It's all about survival, and those who adapt quickly to the circumstances will come out on top. Effective training can be adapted to make use of natural law. Regardless of their designated role, our dogs face tough challenges. We ask them to work with a myriad of distractions, both indoors and outdoors, in changing light conditions, exposed to different sounds, under time pressure, and often under stress. Our task, as trainers, is to ensure the 'survival' of the group, the company, the pack, the team, and the individual. We need to set boundaries, and within

IMPOSING PENALTIES

When we are seeking to change behaviour, punishment should always be considered a last resort. It has tangible dangers; in severe cases, it can damage animals physically and emotionally and it can, and does, damage relationships. Our preference should always be to try to change behaviour through positive reinforcement, but that does not mean ignoring or avoiding problems that could be dangerous to the dogs themselves or to the humans and other animals involved. This basic rule applies not only to dogs but also to people. For example, in my work situation, we do not accept drunkenness at work or the careless use of firearms. We insist that ammunition and weapons are stored with maximum security, and we do not tolerate discriminatory behaviour among colleagues, to name just a few examples.

To enforce these rules, there will, inevitably, be times when we need to make use of penalties. But if this is necessary, it must be done with the best of intentions, honestly, and without emotion or judgment. The punishment must be appropriate, direct, and based on objective data to ensure the behaviour does not return. And, importantly, it can only be practised by a person who knows exactly what they are doing. There are trainers who show weaknesses in timing, decision-making, or poor mechanical skills in the reinforcement process. Such people should never be allowed to work with any form of punishment. If we make mistakes with reinforcement, for example, we reinforce too often, the dog may become overweight. If we under-reinforce, the dog may lose focus, or even leave the training environment. However, if we make mistakes in punishing, we risk and ruin everything. We need to take the utmost care when imposing any form of punishment and, if we lack the necessary skills, knowledge, timing and experience, we should leave well alone!

these limits allow room for error and/or partial autonomy. Crossing borders has consequences. Setting clear boundaries is leadership, as long as it is done with respect, attention, and appreciation.

CORRECTIONS AND THE LEARNING PROCESS

When behaviour crosses boundaries, it can only be stopped or changed by its consequences. Thorndike taught us this simple lesson over a hundred years ago. When the boundaries are clear, the consequences are accepted, understood, and often appreciated. If we ignore a red traffic light, we accept we will get a ticket and a fine. If the amount we pay is sufficiently punitive, it may also change our future behaviour. If we ignore a red traffic light and collide with an on-coming lorry, it will almost certainly change our behaviour and, bearing in mind the *do as I do* principle, the accident may also serve as a warning to other motorists.

Incidentally, there are some interesting cultural differences in this example of the red traffic light. In France, a red traffic light will be ignored at night on an empty intersection. In the Netherlands, drivers will first make sure there is no hidden camera somewhere or a police car and then ignore the traffic lights. And in Germany – even in the middle of the night – drivers will stop in front of an empty intersection and wait for the green light...

Corrections are all around us. Probably the first word a toddler learns is *no!* If the verbal reprimand is ignored and, for example, the toddler goes ahead and touches the hot stove, the consequence – burnt fingers – means the 'touch the stove' behaviour will not be repeated. In addition, the toddler will generalise the meaning of *no!* very quickly. At school, bad grades

First training steps for a camera dog that will progress to being directed via radio.

motivate children to do their homework, being rude to the teacher means being sent out of the classroom and getting into a fight in the playground will result in a detention. We all use corrections, and we all learn from corrections and consequences. *See working with positive and negative consequences, page 125.*

CONDITIONED AVERSIVES

We have already seen that a bridge signal gives the dog positive feedback and is a clear signal that the reward is on its way. When we associate a bridging signal, such as the clicker or a clicker word, e.g., *"good"*, with a reward, such as food or play, it becomes meaningful to the dog and becomes important. In the same way, we can also condition a signal that gives the dog negative feedback. Simply saying *"no"* and linking it to an aversive event, or simply omitting the reward, will condition a punishment signal. A wide variety of terms are used for this, including conditioned aversives, conditioned punishment and delta signals. Negative markers, such as no reward markers (see *page 116),* also fall into this category. Response cost, a term used in human behavioural therapy, is also applicable. This is when a reward (positive reinforcement) is agreed upon and is withdrawn if the unwanted behaviour occurs.

Whatever term we use, it is important to understand that mistakes 'cost' something and have consequences. Ignoring an unwanted behaviour will not change it and may even allow it to become stronger. A clear example of this is the dog who barks at another dog when out on a walk. If his guardian ignores his behaviour, the chances are that the dog will bark more and more and will eventually bark at every dog he meets. In this instance, the use of aversives would not be appropriate because we first need to identify the trigger. If we know what is causing the behaviour, we can work on it. We need to delve deeper instead of arbitrarily using punishment when something happens that we don't like. In the situation of dealing with the barking dog, techniques such as *"look at me"*, or quick changes of direction and speed, or will divert his attention and provide a coping strategy until the cause of his behaviour has been identified and worked on.

Conditioned aversives contain two elements:

- To associate a signal with the withholding of a reward.
- To associate a signal with an aversive event.

Dogs are fast learners, and if the conditioning process has been carried out correctly, they will very quickly avoid or stop behaviours that lead to conditioned aversives. This means that a warning, which could be a verbal cue, a body language signal, or even applying light pressure on the leash, can be used to stop undesired behaviour that would otherwise result in a penalty.

The problem arises if a dog is confused by the signal that is given as a warning. For example, when faced with a conflict situation, a guardian will often call out their dog's name. This is an instinctive response, and no thought is given to the fact that they also use the dog's name as a signal for release, sometimes as an attention signal, and sometimes as a recall cue, followed by a reinforcer. So, we can have a situation when the dog's name is used as a prelude to positive reinforcement and then, moments later, it is used as a warning signal or 'delta' to stop an unwanted behaviour, possibly coupled with a jerk on the leash or another negative consequence. So, we must always ensure that the dog's name does not, unintentionally, become a warning signal – his name must always be associated with good things.

Be careful that your dog's name does not, unintentionally, become a warning signal!

WARNING SIGNALS

Behavioural scientist, Murray Sidman, conducted some experiments on avoidance behaviour in the 1950s that opened up a new perspective on negative reinforcement. The first thing we need to understand is that animals can recognise and avoid events that lead to aversive stimuli. They don't need a trainer to give them a warning signal. In a series of experiments, Sidman found that rats had a precise awareness of time intervals. So, he carried out a series of experiments to see if a warning cue would work like a 'delta' (announcement for aversive) or like a discriminative stimulus (now is the time to do something).

In one of the experiments, the rats had to press down a lever to avoid getting an electric shock. After activating the lever, the next surge of electricity always came after a fixed time interval, in this case 10 seconds. The rats learnt very quickly that they had to press the lever before the 10 seconds were up to avoid the electric shock.

Sidman also brought a red warning light into play, which came on just before the power surge. The rats could now either press the lever before the light went on (the fixed time interval remained as before) or wait until the light came on.

The question was: would the rat interpret the warning light as a conditioned aversive, or as a discriminative stimulus, i.e., where the response is governed by previously reinforced stimulus. The rat has three options:

- Wait for the light (warning signal) to press the lever.
- Ignore the light and keep 'counting the time' and press the lever on the fixed time interval as it used to do.
- Perceive the light as a conditioned aversive, paired with the electric shock.

Assumption 1: The warning light is paired with an aversive and therefore becomes a conditioned aversive itself.
If so, the rat would try to avoid it and pull the lever before the light came on. The fixed time interval remained, and the rat quickly learnt when the light came on.
Assumption 2: The warning light is an important discriminative stimulus and tells the rat what to do – now pull the lever.
If so, the rat would only react when the light came on and would press the lever to avoid the electric shock.

Surprisingly, the rats did not respond until the light came on, so they did not experience it as a conditioned aversive. They delayed their reaction and pushed the lever *after* the warning light came on. So, they considered it to be a discriminative stimulus according to assumption 2.

This gives us some food for thought in relation to warning signals. We can conclude that animals perceive them as discriminatory stimuli, and will probably delay their response until the last moment, just before the aversive occurs. It has been observed that many dogs learn how long they can delay their response, depending on the trainer involved. Animals are also capable of pinpointing warning signals from their environment, even when they do not come directly from the trainer. They learn to associate objects or equipment with aversive elements and will therefore find a way of avoiding collars, leashes, or other objects that have negative associations. This gives us a new perspective on our established beliefs: the warning appears to delay the animal's response.

Let us consider this in relation to an anti-bark collar. When a dog is barking, the collar emits a high-pitch tone, an electric shock or a spray as an aversive. In general, most of these collars will give a warning signal, which starts when a certain threshold is reached. The dog does not stop barking to avoid the warning signal. It's the opposite; the dog will find ways to stay under the threshold to avoid activating the warning signal. This is the moment when dogs start to whine or howl.

For a less extreme use of equipment, let's take a look at a dog on a leash, who is trying to start a fight with another dog. The guardian sees what is about to happen and pulls the leash and prepares to physically stop the dog advancing. The dog keeps hanging on the leash and pulling towards the other dog until the guardian gives a warning – *"no"* – which is connected to an aversive. This halts the dog in his tracks. This was not because he suddenly decided to ignore the other dog or avoid the situation, it was the verbal cue, which, in his mind, heralded an aversive, that changed his behaviour.

Incidentally, we also see the same phenomenon in our own behaviour. When setting a deadline is not sufficient motivation to complete a task, we need to add an aversive element.

Some trainers claim that a warning signal is a conditioned aversive, classically conditioned with an aversive event. For example, if a dog hears *"no, No NO!"* or senses tension on the leash, he will learn about the (unwanted) behaviour he is performing. But is this really the case? What we know is that some animals will stop what they are doing, others will freeze, and others will delay their response. It is interesting to note that the more warning signals the rats received in the experiment, the longer they delayed their reaction to avoid the electric shock. If the warning signal was a conditioned aversive, the rats would have reacted much faster to avoid the situation. It's a bit like reversing in a car: do we slam on the brakes as soon as we hear the first warning beep and jump out of the car to check what is behind us? Or do we continue to reverse slowly and ignore the short beeps until the long beep sounds?

In general, if a warning signal is given some time before the aversive event, we can consider it a discriminative stimulus. On the other hand, if it is followed very quickly by an aversive event, we can consider it a conditioned aversive.

How does this affect the discussion about the use of punishment? Let's take a situation with a strong element of distraction as an example.

A trainer is hired to teach a dog some basic obedience. He trains the dog to sit on a platform and when he takes the exercise outside to the garden, he continues to achieve excellent results. He is just explaining the criteria to the dog's guardian when a cat approaches the platform. She is obviously not afraid and runs straight towards the sitting dog. The trainer knows the dog has been successful in chasing cats in the past so he decides to use punishment because no reinforcer would be strong enough to keep the dog in position. So, as soon as the dog notices the cat, the trainer gives him a verbal warning: *"uh uh uh!"* The dog opts to stay on the platform, the cat walks away, and the dog gets a huge reward.

If the dog had perceived the warning as an aversive stimulus, he would have responded by freezing or would have left the platform. That did not happen. He took the warning as a discriminatory stimulus: *if you stay seated, nothing aversive will happen.* In this scenario, is the warning bad, acceptable, or even good? In fact, it is simply informative and does not make the dog feel uncomfortable. It all depends on the relationship between trainer and dog, previous experiences, trust, consistency, boundaries and consequences.

It is interesting to watch a horse trainer at work, steering 600kg of muscle mass in the desired direction, especially when it is put into situations it would normally avoid. But a well-trained horse will jump over fire, face off against large groups of noisy football hooligans, endure heavy traffic and ignore barking dogs. We should also bear in mind that the horse will not get handfuls of sugar every time it jumps over fire, and I have yet to see a police horse being steered into a group of demonstrators with a carrot under its nose!

If we analyse the process, we can see that horse training is based on negative reinforcement:

- If the rider wants to go to the left, he first puts pressure on the bit in the horse's mouth with the left rein, shifts his weight to the left and gives in the right rein. The horse goes to the left.
- When the rider's thighs exert increased pressure, the horse accelerates, and the pressure eases.

In the hands of a good rider, a well-trained horse responds to extremely subtle signals. However, the basic principle is that the horse has learnt to avoid aversive stimuli by reacting faster and more delicately to given signals, thereby reducing their impact still further. In a top-class partnership, horse and rider move together in perfect harmony and balance, as if they are taking part in a beautiful dance. Which, by the way, gives us a nice parallel: does the lead (male in this case) in a dance couple act aversively when he guides his partner? I would say it all depends on how respectfully he treats his partner, how he attunes to her and how employs the tactics of leadership. If he pushes his partner around, forcing her into changes of direction, she will find his 'leadership' extremely unpleasant. But if he guides her gently, anticipating her movement, she will not find it annoying; on the contrary, she may perceive the guidance as being helpful. Ultimately, whether something is perceived as aversive depends on attitude behind it: the respect, the energy and the intention that is driving it.

Remember, aversives do not have to be negative; it depends on how they are used. They can be helpful if the animal receives information which serves for good. Equally, negative reinforcement does not have to entail the use of a conditioned aversive. It can also be perceived as a warning signal, which allows the animal to delay its reaction.

TRAINING TOOLS

There are many tools that can help us train our dogs: leash, collar, clicker, platform, harness, ball machine, target stick, and many more. These are well-known and widely accepted and are often referred to as the 'whitelist' of training tools. However, there is also a 'blacklist'. Making use of equipment such as a martingale collar or a slip leash, raises eyebrows in some circles, not to mention more extreme measures, such as the e-collar and prong collar, which are banned in many European countries. However, rather than dividing the training community, I would like to address everyone. If we work, train and study together, we will also grow together. Always stay curious and open!

"A fool with a tool is still a fool."

This saying is something we should bear in mind when it comes to animal training. Sadly, there are 'trainers' who only believe in dominance, coercion and punishment, and they continue to find reasons, pretexts, and excuses for using blacklisted training tools. The use of the e-collar and the prong collar may be illegal, but trainers who rely on punishment will always find tools that are not blacklisted. As we know, punishment is easy to administer; it is the consequences which cause the real damage. My aim is to focus on showing guardians/handlers/trainers how to use positive techniques, and to seek expert help if things go wrong. The aim is to improve the quality of life for both dog and human partner, helping them to connect dots, to better understand, to change behaviour and to ensure safety wherever they are, and whatever they are doing.

It is important to understand and acknowledge that we cannot change every type of undesirable behaviour. There are dog lovers who believe that that any unwanted behaviour – no matter how extreme – can be changed simply through positive reinforcement. Unfortunately, this is not the reality. Looking at unwanted behaviour from a human perspective, it is well documented, and ultimately accepted, that people with severe behavioural disorders may need to be restrained, sedated and sometimes excluded from mainstream society. It is no different when dealing with dogs who have severe behavioural issues. In extreme cases, professional trainers can help a dog – and his guardian – to lead an acceptable life, albeit one marked by limitations, instead of resorting to euthanasia.

MOTIVATIONAL AIDS

As is so often the case when discussing punishment, terminology is highly emotive. So instead of punishment, we talk about 'correction' and the tools that are used to correct behaviour are known as 'motivational aids'. The trend of renaming training tools as motivational aids originated in the horse world, where equipment may be used to exert pressure/discomfort as a means of changing behaviour. Motivation sounds like a good thing, but if we take a closer look at these tools within the context of the four quadrants (see *page 59*), they would be classified as:

- Negative reinforcement (R-)

  or
- Positive punishment (P+).

The difference between the two is how the aid is used, and what we want to achieve:

If the aim is to weaken a behaviour, it is a positive punishment (P+).

For example, if a horse was smacked on the nose for attempting to bite, this bad consequence would teach it not to bite.

If we want to reinforce a behaviour, it is a negative reinforcer (R-).

Consider what happens when using a metal bit in a horse's mouth. The reins are attached to the bit so if the rider wants to go to the left, pressure is applied on the left rein. The horse turns its head to the left to avoid the pressure of the bit on its tongue, gums and jaw. As soon as the horse turns its head, the rest of its body follows, and the pressure on the reins ease. A clear example of negative reinforcement.

In addition to the bit, there are other 'motivators' in horse training, such as spurs, often known as 'communication,' aids, which has positive connotations. It would be much simpler if everyone involved in animal training used accurate terminology to describe the principles that are being used: negative reinforcement is a motivator, positive punishment is aversive.

SUMMING UP

We always will be confronted with unwanted and even dangerous behaviour:

- We need to protect our children against aggressive dogs.
- We need to protect other animals from getting killed by dogs.
- We need to make sure that dogs are not the cause of traffic accidents.
- We need to train our dogs to live in our busy society where there is hardly any free space for them.

While we know that behaviour will change by its consequences, we do know that some consequences are much more reinforcing for some dogs than others. For some, chasing a cat in the street is much more reinforcing than the treats he gets from his guardian. However, his chase behaviour is dangerous and risks injury, or even death to the cat, to the dog, and to any road users who may, unwittingly, become involved.

My training is built entirely on using positive reinforcement schedules. But for those unwanted behaviours that can cause severe problems to the dogs in my care, bystanders, road-users or other animals, I will use punishment that is ethical, well thought out, carefully planned and prepared, and executed with a high degree of expertise. The punishment is given without anger or frustration and is well documented so I can reflect on it, change my trainings protocols and learn from it. Ultimately, dogs and humans, are both allowed to make mistakes because we learn from the consequences of our behaviour.

Chapter 18:
TRAIN LIKE A PRO

For me there is no difference between a hobby, sports or a professional trainer. To be effective, trainers of every type need focus, dedication, perseverance, consistency, and a plan to achieve desired objectives. Dogs are so versatile and there are so many potential goals we may want to achieve. For example, our objective may be to:

- Win an important competition or championship.
- Work with a reliable canine partner in operational situations.
- Lead a search and rescue dog in earthquake zones.
- Train guide dogs for the blind.
- Work on conservation tracking programmes.
- Partner a dog trained to indicate illness and contribute to medical research.

In all these roles, we need to ensure that success is made easy for the dog, that their work becomes reliable and consistent, and that they are loved and cared for. Just like us, dogs are social animals, and relationship building is part of the process. From my point of view, anyone who puts the time and energy into training their dog while also working on their own development, deserves to be called a professional.

Where we set the bar is a matter of personal choice, but whatever the demands of the task, we must make sure it is fun and enjoyable, building a sense of trust and co-operation in order to achieve our goals.

BUILDING BLOCKS

I view the different elements of training as building blocks, which means that if we learn how to use them, we can build any behaviour we need from them. A building block is a training element that we can use to build chains of behaviour. For example, I can teach the following training elements as individual building blocks:

- Bridge signal
- Hand signal for 'stop'
- Hand signal for 'sit'
- Hand signal for 'here'

I can then build a chain that looks like this:

Dog walks towards me > stops on hand signal > sits down > waits until I release him again.

Of course, I could also change the order, for example:

'Sit' > 'Here' > 'Stop'

My training blocks are like lego; they come in different sizes, shapes and colours – and they can all be connected. With a bit of imagination and skill, I can put the lego bricks together and build whatever I want out of them, from a small house to an entire lego city. In animal training, I can build chains of behaviour, making use of connections exactly as I would with lego bricks. It's like taking wood, nails, stones and cement and building a house. However, we have to plan the building process otherwise the house will be structurally unsound. For example:

- We cannot put a roof on the house until the walls are finished.

- The walls cannot be built until the foundations are laid.
- We cannot pour concrete on the floor slab until the electricians and plumbers have installed the cables and pipes.
- If we allow the concrete to harden before this work has been undertaken, we will have to prise open the base layer, and risk damaging or weakening the foundations.

House building and, similarly, animal training must start with a plan.

The training module 'hide yourself!' can be very useful in operations with undercover investigations.

When I built my first houses with lego, I placed all the elements that I had already assembled on a base plate. I was then able to build in three dimensions and create a fantasy world. The big picture no longer existed only in my head, it could be seen by others as well. However, this could only happen by investing time and effort in first making a plan, and then seeing it through. A strong foundation is created in this phase of reflection and planning, and so it is with animal training.

Just as we use different sizes and shapes of lego bricks to create houses, cars, aeroplanes or airports, we can also build complex and reliable behaviours by chaining our training building blocks together. This applies equally to training operational dogs, sports dogs, and pet dogs. For example, let's say we start with three basic behaviours – 'sit', 'down' and 'here' – and we train them to a reliable level. Now, using these three training modules, we can start doing more with the dog, taking him to the shops or to a café, for example. We could then add a 'heel' or a more informal 'follow me' and we could move the dog through distracting environments. If we taught him 'take it' and 'leave it', he could pick up dropped items. This process continues depending on the dog's designated role. An agility dog learns directional and handling cues, and how to negotiate each piece of equipment. A police dog may be taught laser cue guidance, bitework, tracking, and how to indicate on target scent. When we follow a good conditioning process, we can assemble all these building blocks in any order. In this way, we can quickly adapt to ever changing circumstances, even in difficult operational situations.

Here, the training modules 'hide yourself' and 'let yourself be carried' were combined – extremely handy for an undercover investigation dog!

It is essential that all the building blocks, which may later represent the links in a complex chain of behaviour, are well trained. Behaviour must be strong and reliable, because if even one link in the chain is weak, the whole chain will break. This is one of the reasons why I train all the links or building blocks separately. I would not attempt to put them into a complex series of behaviours until each element is well conditioned.

THEORETICAL BUILDING BLOCKS

In modern dog training, there are certain keywords and symbols that are repeatedly used by scientists, dog gurus, and other animal trainers. In my own work with trainers, I have noticed, over time, that not everyone uses the same terms or understands them the same way: people start adding information or leaving things out, and we lose clarity. Suddenly, we are no longer talking about training, instead we are having an emotionally charged discussion about the meaning of individual words. Actions speak louder than words, and the ability to train an animal is more important than the words we use to describe what we are doing. However, it is important to use terminology that allows us to identify and follow a common thread in animal training because, regardless of whether we are training a dog, a parrot, a horse, or a marine mammal, many of the training elements are the same.

I am not going to compile a comprehensive list of training terms at this stage, nor get drawn into discussions as to their value. Instead, I want to create a general overview of the most important training basics or building blocks that I use at home, why I use them, and why I believe they are important and relevant. The definitions (in italics) are verbatim quotes from Bob and Marian Bailey, who have had a major influence on my dog training.

What is a stimulus?

A stimulus is any change in physical energy (of many different kinds) to which an organism can respond. Organisms cannot respond to static energy states; a change is necessary. If an organism cannot respond to a given change in energy, it does not constitute a stimulus for it.

The animal must therefore be able to perceive the stimulus: a blind dog cannot see a laser dot, a deaf dog cannot hear the bridge signal, a flashing light is more attractive than a static light. But sometimes we humans need sophisticated devices to enable us to see the change in energy. For example, a laser dot appears to us as a static dot, but a close-up inspection reveals many different elements of light moving around it. It changes all the time, which makes it very attractive to dogs. It may be obvious, but this highlights the importance of checking that the animal we are training can perceive the stimulus we are presenting.

What is a reaction?

A reaction is an arbitrarily selected segment of a behavioural unit. If heeling, for example, is the complete behaviour of the training, the dog's positioning on the trainer's right might be a response in that behaviour.

A pet dog will not respond to the smell of narcotics, but a trained drug detection dog will freeze, bark, lie down, or scratch when he smells them. A patrol dog and a pet dog will react very differently if they are confronted with an intruder. Police dogs will also ignore the sound of gunshot, which would upset most pet dogs. In the course of training, we can change the reaction to a stimulus, or we can strengthen or weaken the reaction to different stimuli.

What is a reflex?

A reflex is a relationship between a stimulus and a response. A reflex does not exist without both its components. It is not a 'thing' in itself, but simply describes what happens when a stimulus occurs and a response follows, under certain conditions and with the certain restrictions that distinguish reflex behaviour from other types.

If a fly zooms towards us, we shut our eyes. This is a reflex reaction, known as the blink reflex. We have the same unconscious response when we salivate if we smell or taste delicious food. We call these reflexes but, in fact, they are only part reflex. What we are describing here is a reaction. A real reflex is the relationship between stimulus and response, such as withdrawing your hand when touching a very hot surface.

What is an operant?

An operant is also a stimulus (or better: a stimulus situation or a stimulus context)-response relationship that requires a further, final component to complete it, namely the environmental effect or the consequence of the stimulus-response combination.

At this point we understand that stimulus-response alone is not the whole truth; there is something else behind it. This involves understanding the ABC of behaviour: the Antecedents, the Behaviour and its Consequences (see *page 86*). Skinner chose the term 'operant' because it 'operates' on the environment. In many cases we see that it is not just the stimulus that triggers the response, there is more going on. The black box (see *page 124)* is often referenced in this context. When a stimulus occurs, the information in the black box (the animal's internal thinking) is processed and we see a reaction. Here the environment (the context) will determine the behaviour.

For example, if the traffic light *(stimulus)* is green and the road in front is clear *(context/environment)*, the driver moves off *(reaction)*. The reaction is reinforced by the consequence of moving forward *(reinforcer)*. If the driver fails to check that the road is clear, they may cause an accident; if they accelerate when the light changes to green, they get a speeding ticket. Both the accident and the ticket can affect the meaning of the antecedent *(green light)*. The green traffic light can even become a warning signal for an accident or a speeding ticket. This is how conditioning works.

Internal behaviour

Some psychologists classify yet another type of nervous system behaviour, event, or activity that we call internal, private, or mental, such as thinking, planning, imagining, or feeling. While we take it for granted that our own ideas, thoughts and feelings are real, it can be very difficult to describe them to other people. We cannot directly know what someone else is thinking or feeling. Many also relate to their dogs' thinking, feeling, planning, etc. We don't deny that animals think, feel, and even plan. However, we still lack the objective possibility of checking that dogs are really involved in these activities, and in which ones. After all, you have no way of knowing what your child, spouse or best friend is thinking – how on earth can you offer anything other than speculation about your dog's mental activities?

We are currently learning more about this exciting topic through inventions such magnetic resonance imaging (MRI) and computerised tomography (CT). In his book, *How Dogs Love Us*, American neuroscientist, Gregory Berns, describes what he has discovered through using these technologies. The brains of dogs were examined using MRI as they were presented with previously conditioned signals, such as hand signals. A hand straight up meant food (hot-dog) is coming; two hands vertical pointing at each other meant no food is coming. In this simple way, the MRI scanner could spot activity in the parts of the brain where the reward system is located. This has allowed scientists to study the reward system response in relation to other stimuli, such as presenting the smell or a photo of the dog's guardian.

We cannot simplify internal behaviour to the point where we always expect a given signal to be followed by the same reaction. For example, the cue we use for 'sit' is not always followed by a perfect, quick sit (reaction). Sometimes the dog sits down more slowly than expected. Sometimes we get a sloppy side seat on one hip, and sometimes we even have to repeat *"sit"* a couple of times in order to get a response.

To view it from a human perspective, let's consider the example of a Friday night trip to the cinema. What triggers us to go? Was it the TV commercial for the film? An advert in a magazine? A trailer on social media? Is the stimulus that makes us go to the movies always the same? Our emotions trigger actions or influence our decisions; they impact on how we deal with situations and how we act.

- When we see an image of a dog that appeals to us, we struggle to stay emotionally objective.
- When we interact with a baby, we not only change our body language, but also our voice.
- When a large, unfamiliar dog runs up to us and fixes us with a hard stare, we feel less comfortable than when a small dog approaches, wagging his tail.

In addition, we cannot get away from being anthropomorphic, ascribing human characteristics, feelings or intentions to our dogs. Again, we have no objective way of finding out what our dogs are really feeling, thinking, or wanting. Once we start seeing their body language and actions through our human emotional filter, it's difficult to remain objective. But that is exactly what trainers need to do. When we are working with dogs we need the objectivity of a scientist, which is why we collect data, work double blind, and use criteria to decide whether to reinforce or not. Instead of making judgements and decisions based on feelings, we need to use facts.

We can only speculate as to what our dogs are thinking about.

Chapter 19:
TARGET TRAINING

In general, target training means that the animal, in this case a dog, is sent to a target and touches it with his nose. The contact leads to a bridge signal and reinforcement, or the dog gets a signal to do something near the target. Target training is not only fun, it is also one of the basics in my training plans. Targets can be used to evoke specific behaviour. For example, I can signal the dog to touch, look, place or pick up something (and of course many other behaviours). For dogs, targets are like our post-it notes, they guide the dog to the right position and tell him what to do there. They also increase success rates because, with the help of targets, the dog can better understand what we want from him.

The 'classic' in target training: the dog touches a target with his nose.

Target training is the perfect way to teach a dog to go to a specific place or object. We can also use targets to train dogs to complete an obstacle course, because once he knows how to touch a target, a series of targets can be placed

Line-up.

on all sorts of obstacles. I like to use a platform as a start and train the dog to run from there to a visual target. This creates a procedure, which means that any obstacle can now be placed between the start and the target. The training is easy to understand and because there is a high rate of success, the dog is encouraged to run from the start to the target using the shortest possible route. He will jump, crawl, climb stairs, walk over a footbridge or an inclined wall, run a slalom or, by-pass any other obstacle that lies between the start and the target. Targets do not necessarily have to be nose targets. We can easily train a dog to stand on a target, or touch a target with his front paws, which could be used to train him to run all the way down a sloping wall instead of jumping off it before the end. Targets are very powerful!

In scent detection work we use line-ups and scent wheels as targets, and the dogs quickly learn that the containers are worth investigating. They know that, in all probability, they will find target scent in one of these 'targets' which will lead to a reward.

From these targets we can quickly generalise to searches outside the comfort of a scent wheel or line-up. To start this process, we place the equipment in different positions within the room and change the environment.

Target training gives us the opportunity to work a lot with positive reinforcement. The criteria of reaching the target and carrying out a specific behaviour, such as touching it with the nose or dropping an electronic device on that exact spot, is crystal clear to both dog and trainer, and therefore success rate – and reinforcement – is high. A well-conditioned target acts like a magnet for dogs and can be used to set them on the path to error-free learning.

Another great advantage of target training is that we can easily switch to

A scent wheel gives the dog a variety of containers to investigate.

When working with a line-up, the containers are moved from set positions.

In this way, the dog learns the 'sniffing game' in small steps and we can put him on the road to success.

RIGHT ON TARGET!

I became convinced of the benefits of targets when I designed the training programme for radio directional camera dogs. When I started work on this in 1996, I was on my own – no one had attempted to do this before. There were a handful of camera dogs around the world, but they could not be remotely controlled with directional commands. They were simply search dogs carrying around a bulky camera system. Attempts by trainers to steer them in any direction relied heavily on coercion: the dogs were punished when they went in the wrong direction and punished when they dared not move.

My plan was to use targets to build up an expectation that something good that was going to happen out there. By teaching the dogs to run to targets, and using positive reinforcement, bridge signals, and varied training schedules, I was able to teach directional commands that I gave them over the radio. My dogs loved this game! Soon they were scanning the area, not for a rabbit they might be able to chase, but for a target to run to, because they knew they would find a reinforcer in the form of a feeder or a ball launcher. I also arranged it so that the reinforcer was not available at the precise location of the target; the dogs had to wait for another directional command to access it. The dogs quickly understood to go left, right, forward, come back and much more on a signal – i.e., a voice over the radio.

Since my first radio directional camera dog training programme in 1996, I have trained many special forces around the world, and it has also been copied by trainers from many disciplines, proving the power of target training.

Targets are an integral part in the training of our radio-controlled directional camera dogs.

Targets will help the dogs to understand directional commands very quickly. In addition, they build up high motivation and expectation: where is my target?

other environments. If we use familiar targets when we start working in a new open space or in a new building, the dog will generalise very quickly. The targets give him confidence, and also contribute to desensitisation with regard to environmental stimuli. They can also be used when we are teaching a dog to switch from a trained object to something that is unfamiliar. For example, we could spend weeks training dogs to place a specific device on a certain car model. But when the car was no longer a Mercedes but a VW, we had to start training from scratch. However, when we started using a specific target, we were able to switch from one car model to the next in just a few repetitions.

CHOOSING TARGETS

The targets I have described are fixed; they do not move. Each target is placed in position and is used to 'station' an animal. Once the animal reaches the target, it gets a reinforcer or finds a further cue. The basic goal is for the animal to stay at its station for a shorter or longer time.

- Dolphins wait at stationary targets while a selected dolphin performs a trick.
- Birds fly to stationary targets and stay there until recalled by the falconer.

In dog training, we make good use of a platform as a stationary target, teaching the dog to stay in position while their guardian is busy, or when a trainer is setting up equipment. The dog is placed on the platform and given the same predetermined command – e.g., *"sit"* or *"down"* – so he knows what to do there, namely stay there until he receives a further instruction. The criteria relating to the platform are therefore very clear for him. The handler can set up the training props, or watch others train while their dog waits nearby. In this way, the dog is involved in what is happening which, in my opinion, is better than being excluded and waiting in the back of a car.

The platform serves as a stationary target and starting point for Charlie. He waits there while I prepare a line-up with scent samples.

BROADER APPLICATIONS

The next time you attend a dolphin show or a show involving other marine mammals, take a closer look at how targets are used:

- Large targets are attached to long poles and are used by the trainers to manipulate behaviour.
- Balls are attached to ropes, suspended from the ceiling, to encourage the dolphins to jump high out of the water.
- The platforms where the dolphins gather are fitted with targets, and the trainers use hand target signals to direct the dolphins to certain positions.

A variety of targets are used in dolphin training.

In the Netherlands there are some famous shows featuring sea lions. They are very entertaining to watch, but even more interesting when viewed from a trainer's perspective. The sea lions act out entire stories, travel specific routes, open and close doors or windows, ring a bell, pick up items and interact with the trainers. There are many 'hidden' targets to give directions and signals to the animals.

Birds of prey shows also make clever use of targets. The birds are conditioned to fly from their starting boxes to their stations, where they receive a reinforcer or another signal to fly to the next station. These stations or targets are hidden in the show grounds. For example, it could be a camouflaged box hung in a tree, a wooden pole, a rock, or the falconer's gloved arm. The next time you attend a birds of prey show, try counting all the targets in play!

Birds of prey can be taught to fly to a target, which takes the form of a station.

DYNAMIC TARGETS
Dynamic targets move! A dynamic target can change position, height, space or environment. This can be a target or a person that the animal should follow. For example, in free-flying bird shows, the birds would be trained to follow the person in a red cap. A large bird of prey has also been trained to catch a drone from mid-air, thereby preventing unauthorised filming relating to drug smuggling.

When I first started training my camera dogs in directional cues, I used small flags as targets, which we moved to get the dog's attention. We have now graduated to lasers, which are the top end of dynamic targets. They can be used to send an explosives dog to search a car, or to direct him from one car to the next. I have also trained police intervention team dogs to attack a person, marked with a laser dot on their chest.

Generally, the targets we use should be as simple as possible, easy to replace and readily available. In my operational role, I travelled around the world and it was important to get hold of the same targets no matter where I was, even if it meant getting hold of the material so I could create my own. Targets need to be constructed in such a way that the criteria are very clear, not only for the dog but also for the trainer.

FADING TARGETS
In marine mammal shows, free-flying bird shows and many other training programmes, targets can continue to be used and there is no need to phase them out. At the start of training, the targets are large and easy to see, smell or hear. As training progresses, distractions are added – other animals, machines, sounds or people – while the big target is in place. It serves as a focus

With a disc-shaped target (pictured) the task of touching the target is difficult for the dog and that makes it hard for the trainer to time the bridge signal exactly. If the trainer reinforces too soon, the dog will stop short of the target and not touch it; if the reinforcement is too late, the dog will push too hard.

With a different type of target, the training becomes much clearer for both dog and trainer. The dog has to stick his nose into the target so the criterion, 'nose in the target', is the moment to give the bridge signal/reinforcer. Here, an automatic feeder is positioned next to the target so the target behaviour, 'nose in target', can be reinforced very quickly with a bridge signal and with food close by.

for the animal, promotes confidence and 'tells' it what to do. As soon as the animal works reliably in a distracting environment, the targets can be made smaller. In time, the show is structured in such a way that audience can hardly see the targets, and they are unaware of the subtle signals given by the trainers or the small, hidden targets. I really enjoy working with targets because they give the animal so much information and security. So, given the choice, if it is ok to work with targets, no matter how small, I would always use them.

Of course, we cannot work with targets in operational missions – I cannot prepare the scene beforehand by placing targets. Imagine if a member of a special task force had to put himself in a dangerous situation to deploy targets before sending his dog to find them. Therefore, when I am working with operational dogs, I fade out targets very quickly. Once the dog is working reliably, I gradually reduce the size of the target until it eventually disappears entirely and is replaced by a stimulus picture. For example, in the case of my radio-controlled camera dogs, I would use a door lock. Door locks are everywhere, inside and outside and in all countries. All

The target on the door lock gets smaller and smaller each time until it disappears completely. Reinforcement comes from the automatic feeder (left).

I needed to do was place a clearly visible target on a door lock at the beginning of training and gradually make it smaller and smaller. This creates a strong stimulus picture (door lock) which will trigger a reliable response.

I also fade out the body language cues that the camera dogs learn, so that we can control them remotely, via radio, with the help of targets. We do this by setting up a series of targets in a small space and sending the dogs in a targeted manner: 90 degrees right, 90 degrees left, straight ahead, and so on.

Extreme target training, with many targets and different directional command, is one of the elements in our training of radio-controlled directional camera dogs.

Chapter 20:
WORKING WITH TRAINING PROTOCOLS

Training is about learning new skills and competencies and reaching a level of fluency so as to execute them, without thinking, at the highest level. My aim is to train fast, fluid, reliable and strong behaviour. However, it is often the case that trainers structure exercises in such a way that it is easy to get the desired results. Of course, we want to make success easy for the dog but if we want to learn and achieve more, we need test ourselves, and our dogs, and keep growing. We need to step out of our comfort zone.

When I started working with lasers as targets in 1996, I didn't have any training protocols. I just did what seemed right to me. This was a time when training techniques were applied randomly; there was no science, no data, and no verification. We just kept trying. Working with Bob and Marian Bailey changed all this. They taught me that collecting data was the means of designing effective training plans. My training became deterministic, which meant that I started fact-checking everything. I recorded all my findings, and the benefits were enormous.

Back then, a laser unit cost $25,000 and we had to build it from individual components. It took us six to eight months to teach our dogs to walk 200m from the starting point to the laser point to complete a task. But once I started collecting and analysing data, I was able to write protocols – and save training time. I became more efficient and was able to train more dogs. Eventually, I came up with a plan that allowed us to train laser control in just three weeks.

Nowadays trainers often share their 'progress logs' and other simple training plans. Good to see the world of dog training is changing! The fact that trainers are starting to work from designated plans is a big step forward – and even a small and simple plan is better than training ad hoc. When we make it up as we go along, we tend to forget the weaker moments and the worst results. In contrast, a plan enables us to take small steps to reach our end goal. My protocols are therefore much more than just progress logs: they have clear instructions and very clear criteria for both dog and trainer. There is no room for guesswork or error – even the smallest details are important.

WHAT IS A TRAINING PROTOCOL?

A training protocol provides a fixed procedure, which is generally the result of much trial and error. We need to be certain that we have found the best way of training a behaviour (or behaviours) before including it in a training protocol. In addition, we need to factor in distractions; in my case, this would be the sound of gunshot, helicopters or explosions.

I write a plan and follow it, which means undertaking a number of sessions and trials, and evaluating the results. When I have found the most efficient and effective method of achieving my goal, I write it into the training protocol, which means it can be used on future occasions.

So, it's not just about logging the results of a training session. We need to think and plan ahead; we need to work on a protocol, and then stick to it.

A training plan is the result of data collection, which can only happen if training sessions are made up of carefully planned trials. I plan a series of sessions, each with a designated number of trials. Then, when I analyse the data, I can develop a protocol which will become the procedure for future training.

Let's take the example of a log for a 'sit' protocol. In this case it is for a more advanced dog, rather than training the initial stages. My entire 'sit' workout consists of 25 sessions, each with 10 trials. Only when the dog has been successful in at least 8 out of 10 trials will I move on to the next session.

Looking at the example (*below*), we have already built in quite a bit of variation. The distance from dog to trainer is set in advance for each trial, and the length of time the dog should sit is also planned in advance:

- The trainer cues the dog to sit, drops the leash, and walks the distance specified in the plan.
- Once in position, the trainer stops and looks at the dog for the duration stipulated in the protocol.
- After the trainer gives the bridge signal, the dog is moves quickly towards him/her to receive the reinforcement.

Following this example, you can create your own plan.

Working with training plans and protocols allows us to get out of our comfort zone and grow – not only as trainers, but as partnerships. Plans and protocols focus on both dogs and humans. The process of training is about decision-making, and this is where the plans and protocols can help. They tell us when to move on to the next step and guide us to a higher level of training. In this way, we will be able to achieve the goals of our dreams!

Trial	Distance between dog and trainer	Sit duration in seconds	Bridge signal	Reset	Remarks
01	10 metres	30	X		
02	25 metres	10		X	
03	50 metres	40	X		
04	15 metres	15	X		
05	30 metres	60		X	Cat walking past
06	80 metres	05	X		
07	50 metres	30	X		
08	40 metres	20	X		
09	50 metres	90	X		
10	20 metres	30		X	Child running past
			70%	30%	

This session, consisting of 10 trials, is midway through the 'sit' protocol.

Marieke and Ygor in trial no. 4 of the 'sit' protocol.

Training protocols can be likened to a manual for training a certain behaviour. They help us to make decisions, describe the steps we need to take, and predict the outcome of the training process. The criteria, such as distance, equipment, body language, position, environment, distractions, reinforcers, number of repetitions, etc., are described in detail. Training protocols must be written with clarity so that they can be read and understood by everyone – even those with no experience of animal training. This can be facilitated by photos, diagrams or even video clips to explain exactly what we want to achieve.

Training protocols are not set in stone, but should be viewed as living documents. When we make an improvement, or understand something better in the training process, we add it to our protocol, which becomes better, more effective and more efficient. The process of making structured plans, collecting data, and incorporating our experiences, cuts training time, and improves outcomes.

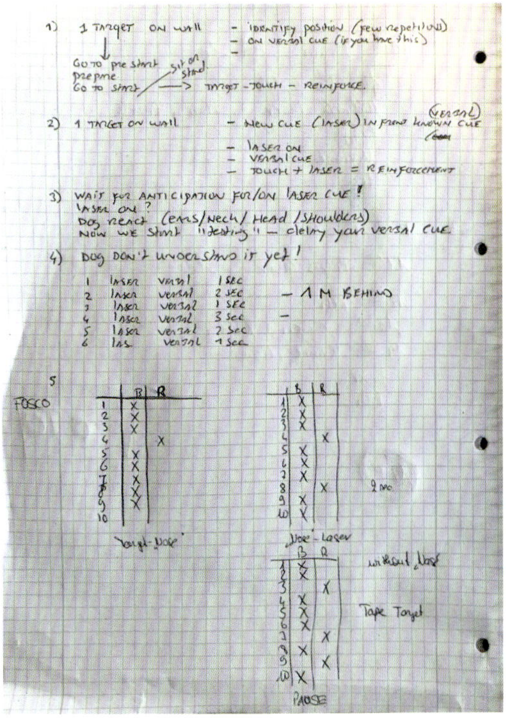

1) 1 TARGET ON WALL – IDENTIFY POSITION (FEW REPETITIONS)
– ON VERBAL CUE (IF YOU HAVE THIS)
–
GO TO PRE START / PREPARE / GO TO START (SIT OR STAND) ⟶ TARGET – TOUCH – REINFORCE.

2) 1 TARGET ON WALL – NEW CUE (LASER) IN FRONT KNOWN CUE (VERBAL)
– LASER ON
– VERBAL CUE
– TOUCH + LASER = REINFORCEMENT

3) WAIT FOR ANTICIPATION FOR/ON LASER CUE!
LASER ON?
DOG REACT (EARS/NECK/HEAD/SHOULDERS)
NOW WE START "fading" – delay your verbal cue

4) DOG DON'T UNDERSTAND IT YET!

1	LASER	VERBAL	1 SEC	
2	LASER	VERBAL	2 SEC	– 1 M BEHIND
3	LASER	VERBAL	1 SEC	
4	LASER	VERBAL	3 SEC	–
5	LASER	VERBAL	2 SEC	
6	LAS	VERBAL	1 SEC	

5

FOSCO

	B	R
1	X	
2	X	
3	X	
4		X
5	X	
6	X	
7	X	
8	X	
9	X	
10		

'Target-Nose'

	B	R
1	X	
2	X	
3	X	
4		X
5	X	
6	X	
7	X	
8		X
9	X	
10	X	

2 m

'Nose' – Laser

	B	R
1	X	
2	X	
3		X
4	X	
5	X	
6	X	
7		X
8	X	
9		X
10	X	

without 'Nose'

Tape Target

PAUSE

An example of a training protocol with data collection.

THE ELEMENTS

Creating a protocol is like building a house: there are a number of basic elements to any house, large or small, such as the foundations, the plumbing, electrical wiring, doors and windows, roof, walls, living room, kitchen, bedroom and so on. There are also basic elements in our training protocols, which allow us to develop new behaviours to fit our purpose and to adapt them, at speed, to meet changing circumstances.

Bear in mind, a dog is not a computer where we can swap the hard drive and change the operating system. Once a dog has learnt a new skill, new scent or new signal, it's difficult to erase that information. We always get what we reinforce – so we must only reinforce what we want. We need to be smart about what we teach, in what order, and how best to maintain it.

KNOWLEDGE AND SKILLS

Before we start training, we need to be sure that we are at the right entry level. We tend to focus on the dog's aptitude for learning a new skill, but the trainer's knowledge and skill is equally important. When training a dog and handler, we need to pay attention not only to their strong behaviours but also to their shortcomings. I have worked with trainers who were convinced they knew how to use a clicker but, in fact, they had no idea how to use a bridge signal, or why they needed one. I have also met trainers who refused to reinforce with anything other than a tennis ball because they had been told that it was the best reward for all dogs.

The key to success is to start a training plan with a good dog and a good trainer. Test not only the dog, but also the handler their knowledge, their willingness to learn and their mechanical skills. Does the handler have:

- The necessary hand-eye co-ordination to clicker train effectively?
- The ability to translate theory to practice?
- The ability – and willingness – to follow instructions?
- The capacity for self-reflection?
- And above all, the willingness to be coached and to change their behaviour where necessary?

GOAL SETTING

It is important to set an accurate and realistic goal. The vague goal of training a tracking dog, a scent detection dog or an attack dog will not do. We need to be more specific. What exactly do we want to achieve?

- A tracking dog that can:
- Follow a track on a hard surface, at night, in a distraction-free environment?
- Follow a track through a village during the day?
- Or maybe track through the centre of a big city?
- A scent detection dog that can:
- Identify 10 different scents?
- Find target scent in a line-up of different odours?
- Identify minimal amounts of target scent?

We need to decide on our goals and write them down in a way that everyone can understand – even those who have no idea of a dog's capabilities – which was something I came up against when trying to convince a unit commander (it could be a manager, or a medical director) to

Goals must be clearly defined.

fund a training programme. We also need to verify that we have met our goals and reached the required standard, which could mean handing out the training data, or conducting a specified test.

IT'S ALL ABOUT CRITERIA!

Goals are built from criteria, and their importance should never be underestimated. Watching a football game without lines, goals and rules is not fun. Criteria are the parameters by which we train our dogs and check their work. The steps within the criteria give us the direction to follow, the progression, and the timescale. The criteria must be sharp, clear and consistent:

- A 10-minute track on asphalt in the city is not the same as an 800m track in the woods.
- We cannot teach a scent detection dog to find a minimum of 5 grams of cocaine while working through the training phases using 100 grams.
- A two-minute sit is different from a 10-second sit.
- A police dog trained to attack a screaming 'suspect' in a protective bite-suit will not attack a 'silent' criminal dressed in jeans and a t-shirt.

As already outlined, my training plan for 'sit' consists of 25 sessions, each incorporating a set number of trials. Session 01 starts with very easy criteria which get more difficult as we progress. If the dog initially works at a 5m distance from his trainer, session 25 will entail the trainer being positioned at a distance of 50m, out of sight, for a duration of five minutes. Criteria builds expectation and reliable behaviour. However, this can only be achieved if the instructions are clear and progression is incremental. I work in small steps and only incorporate criteria that I can describe in a maximum of five words – *'dog sits on platform'*, *'dog touches target with nose'*, *'touch target on signal'*. If the criteria require more than five words of explanation, the step is too big.

SESSIONS AND TRIALS

The training sessions and trials that are used to design a protocol need careful planning. Let's revisit my protocol for 'sit,' which extends over 25 sessions. One of these sessions will take place in a closed environment with no distractions, another might be in front of the entrance of a busy supermarket with lots of distractions, or maybe in a field with rabbits. Each session has a set number of trials; I generally plan an average 10 trials. This allows me to measure the effectiveness of the session based on a percentage. If a dog is successful in 80 per cent of trials, i.e., eight out of 10 trials, I rate the session as effective. If this is repeated in the next session, i.e., the dog has achieved a success rate of 80 per cent or more in both sessions, we are ready to progress. In this way, I know for certain that the dog has been successful in a minimum of 16 trials, over two sessions, before increasing the challenge. With these results, I am satisfied that the aspect of behaviour I have been working on is well conditioned, and we are ready to move on to the next step, which could be an increase in duration, distance, or distractions.

If we always build our training protocols with a progression of 80 per cent success over two sessions, we can collect and analyse data and measure progress. If we are failing to meet the desired success rate, we may need to change the plan. Remember, a protocol is a living document and can always be changed and adapted.

REINFORCEMENT

Positive reinforcement is very powerful, not only for the animal but also for the trainer. As previously highlighted, trainers need reinforcement, too (see *page 149*). Some of our training programmes stretch over months and if the trainers only get reinforcements at the end, the dry spell can be lengthy. For this reason, I conduct intermediate exams or evaluation tests. In my experience, trainers like to show off the results of their work, and it keeps them focused on their goal. It's also a good opportunity for giving feedback – and it's the perfect moment to give and receive reinforcement.

BOUNDARY SETTING

As I have already emphasised, boundaries are an important aspect of training. If they are exceeded, there are consequences. Within these limits, however, there is a great deal of autonomy for both dog and trainer. To make this clear, we need criteria so that everyone understands the extent of their autonomy. In this way, a safe learning environment can be created; a place where the dog and trainer are allowed to make mistakes in order to learn from them, a place where the trainers are not afraid to ask for help. There is also no room for frustration or aggression within these boundaries. We can hold each other accountable for what we do. Again, our training schedules and how we communicate with each other is highly influential.

PLANNING AND PREPARATION

Designing a successful training protocol involves careful planning and preparation, considering all the elements in detail, and progressing in small steps. These are the factors that need to be considered:

OPENING UP COMMUNICATION CHANNELS

We will be teaching different cues/signals and conditioning various bridging signals, such as a clicker, whistle and a marker word. We, therefore, need to start by creating a reliable communication system, bearing in mind the Clever Hans principle (see *page 131*). In addition to verbal cues and bridge signals, we will also be using body language in training. The exception would be if the end goal was for the dog to work with the handler out of sight, in which case building body language into the training plan is irrelevant.

Communication is a two-way street, so we also need to learn how to read the dog's body language:

- How does he show if he is feeling ok, happy or sad?

- What exactly do we see when he finds target scent?
- How does he react when he has found or lost the track?
- How can we tell if he is feeling nervous, anxious or aggressive?
- What are the triggers that make him go from one state of mind to another?
- How can we tell if the dog has noticed something that we cannot detect, such as a specific smell or a high-pitched sound?

Clear communication is not only achieved through training, it involves spending quality time together, which could mean enjoying a quiet evening at home or going for walks and observing how the dog reacts to sounds, smells, movement, and other distractions. It's all about being together, doing things together, and discovering things together.

THE LEARNING ENVIRONMENT

I begin training in a distraction-free environment. I want to build positive expectations and trust – and show the dog that I am the greatest! Anything and everything is triggered by me. Rewards come from me. Games are initiated by me. Help comes from me. In an environment free of distractions, the dog learns very quickly to keep an eye on me, because I create exciting things! For this reason, I start in a completely empty, boring room – no other people, animals or even furniture. Just the dog and me. In this way, I am confident that I am in control of the environment. This is not possible in an outside setting where a bird may fly by, a rabbit may run across the field, or a cat may appear on the training ground. In addition, when I am training scent detection dogs, I can control the airflow and temperature indoors – and I can be confident that it will not rain. This is how I make it easier for my dogs to succeed.

Following a series of repetitions working in a closed environment – when I have built strong behaviours, gained

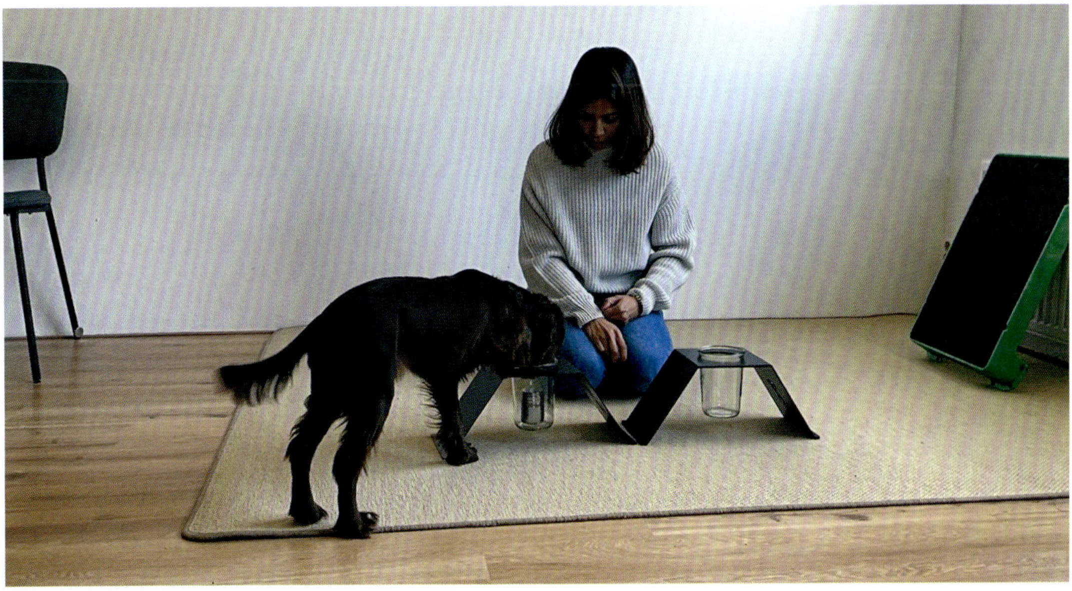

Start in a closed training environment with no distractions.

some fluency, and consistency – I start to introduce distractions. I move into larger spaces, but still with minimal distractions, and progress slowly until I can finally work outside in a world full of distractions. Collecting and analysing data will tell us when it's time to change the environment.

CHOOSING A BRIDGE SIGNAL

The use of a bridge signal is fundamental to my style of training, and I use a variety. Generally speaking, a bridge signal means: *'Well done, a reward is coming'*. For my dogs, a whistle means: *'Come back to the trainer to get your reward'*, while the clicker tells them: *'The trainer will come to you and bring you a reward'*. I also use a ball machine or a feeder, with built-in bridge signals, that I hide in the training environment.

The choice of which bridge signal I use is based on the end goal, i.e., the situation in which the dog will be working. For example, if I am working undercover, using a loud whistle is out of the question but I could use a high-frequency whistle, which only the dog can hear. Equally, if the dog is working at a great distance from me, I will not use a clicker as the dog won't hear it. In this case, a whistle would be the better choice. Remember, whichever bridge signal is selected, the dog must be able to hear it, and it must be applicable to future working environments.

In addition to choosing from a clicker, a whistle, an ultrasonic cue, a radio transmission, tactile contact, or any other bridging signal, we also need to condition a bridging or marker word. I have lost count of the times when this has saved me on missions when, for some reason, I was unable to use my usual bridge signal, or the device was broken or lost.

PROGRESSIVE TRAINING

We need to break the behaviour we are training into small steps – we must learn to crawl before we learn to walk. Again, errorless learning comes into play (see *page 123*). If we make the steps sufficiently small, the dog cannot help but make the right decision and will succeed every time. However, this type of learning comes at a price. Taking tiny steps is time consuming and, at some point, the dog must learn that to achieve success, he may have to overcome obstacles on the way. If things come too easily, and we never have to work hard to get results, we don't learn perseverance – and it is just the same with dogs.

Communication and observation are important here. By now we know how to communicate with the dog, and we can see if he is comfortable or stressed. We also know which reinforcers he really likes. This allows us to start with simple tasks and to gradually increase the challenge, working on distance, duration, distractions, and degree of difficulty. Remember, the dog will always be working on the *'what's in for me'* principle (see *page 49*) and a *'sit'* in a distraction-free training room is different from a *'sit'* at the entrance to a supermarket. Remaining seated in a highly distracting environment needs to be 'paid for' with a higher quality reinforcer and built up in small, achievable steps.

GIVING REINFORCEMENT

The trainer is not the only person who can provide reinforcement. There are times when we can make use of a third party or a piece of equipment, such as an automatic feeder or ball throwing machine. Gundog trainers also make use of dummy throwing machines. However, we need to make sure that this procedure fits the ultimate goal. For example, a patrol dog should not get used to bystanders rewarding him during training. Imagine what would happen if he was tasked with keeping football hooligans at bay and someone threw a bag, or some other item, which he thinks he should investigate in the hope of finding a reward?

In whatever way the reinforcement is delivered – by the trainer, by a third party or via a piece of equipment – it should come as a surprise. The dog should always be questioning the *what, when, where* and *how* because the reinforcer is at its best when it comes unexpectedly. In the initial stages of training, we reward quickly and close to the target or the source of the scent. But after just a few repetitions, we can go further and slow down a bit. The bridge signal lets the dog know that the reward is about to come. In the meantime, we also know that it is not so important *what* we give the dog, but *how* we give it. With more experienced dogs, I make a little show of it by putting my hand in my pocket and, very slowly and mysteriously, pulling out an awesome reward. Or, if I am using a ball machine, I make it seem as though the ball falls out of the sky, or, I allow a dog to stumble across an automatic feeder hidden in a bush. In all these scenarios, we still need to keep an eye on the reinforcement rate, collect data, and remember the reinforcement schedules.

I also like using the technique of tossing several pieces of food in a row (see *page* 97) or, alternatively, using automatic feeders. The advantage of throwing food is that we can achieve a high rate of reinforcement. Of course, the trainer must have a good aim and be proficient at throwing at speed, but once this technique is mastered, we can reward the dog at a high rate, even from a distance. This is useful in platform training, for example, where we can easily reinforce 'sit', 'down' or 'stand' on the platform by throwing food several times in a row. The advantage is that we don't interrupt or disturb the behaviour by having to go to the dog after the bridge signal to deliver the food. By throwing food on to the platform, we turn it into a kind of magnet for the dog and thus, it becomes a powerful reinforcer. We can use this same technique in the park or in very distracting environments so that, even if we are at some distance from the dog, we are the source of multiple reinforcers. Automatic feeders can be used in a similar way; they allow us to work at a distance, or even out of a sight – a very powerful training tool!

FIRST STEPS

Following planning and preparation, the first practical step is to 'get' the behaviour. At this stage, we work with luring and shaping to train the desired behaviour quickly. For example, we would establish a *'sit'* by holding a hand with food over the dog's head and moving it backwards a little. The dog will then, almost automatically, assume the *'sit'* position. If we were working on a *'down'*, we would briefly present food in the hand between the dog's front paws and then move it towards his belly. The dog lies down, the hand opens, and he is reinforced. In these situations, the dog finds out very quickly which behaviour, or which body position, gets reinforcement. As soon as the dog begins to anticipate our body language signals, we can put the new behaviour on cue.

WORKING WITH DATA

So now we have invested time in thinking and planning, we are ready to activate the training plan and protocols. Some trainers find it difficult to stick to their plans. They are too emotional, too influenced by mistakes that occur, or just not as disciplined as an animal trainer should be. We need to work with our plan – it's the only way to collect data and to fine-tune training techniques. When something goes wrong in training, it will be easy to see it in the data as the rate of reinforcement goes down, and the success ratio may decrease. This allows us to reflect on what might have happened. Was there a distraction that we didn't notice? Was the value of the reinforcer sufficient for the task? By checking the data, we can adjust the protocol for the next session, if necessary. This not only helps the individual dog, but also other dogs that may be taught using the protocol in the future. It will also help fellow trainers and friends to make better plans!

When I started designing training protocols, I wrestled with finding the best system for collecting data. In fact, I have changed it several times since then and will probably change it several more times in the future. `I have discovered that there is only one way to find the ideal system for collecting data – just start! In 1996 I began by writing in notebooks, which led to many files with plans and pages full of notes. Fortunately, today we work with computers and smartphones which are ideal for the speedy collection of data. There are many apps and programmes to help with this; most will automatically generate the date, time, weather, location, etc., so we can focus on the protocol itself, noting the session, the number of trials, the success rate and the distractions.

With the collected data we can analyse what is really going on, as it forces us to be objective. Nature has programmed us to forget the bad things and it is all too easy to fool ourselves. For example, we may be feeling good after completing a session of 10-15 trials but when we look at the data, we find that the criteria have not been met in every trial. This is not a major concern – it happens to the best of us. However, it teaches us that the only reliable information comes from the data – not on how we feel. We should never change a protocol based on our emotional response; we must make decisions based on the data. If the rate of reinforcement is too low, we need to change the criteria to make the task easier or add intermediate steps. For example, we could decrease the distance, the duration, the weight the dog has to carry; we may need to offer the target scent in a higher concentration or move to a less distracting environment.

First get the behaviour, then introduce the cue.

When we are training, there are a few golden rules that always apply. Conditioning the behaviour *first* and *then* introducing the signal is one. Novice trainers often make the mistake of giving the dog a new command – which means nothing to him – and then attempting to condition the new behaviour. This does work, but it adds an unnecessary element of confusion. The dog will learn a new behaviour much more effectively if we work without distractions and extra information, such as body language. But that is very hard to achieve because we will always, unintentionally, give away clues. This extra information, along with, sounds, smells, and context, becomes important for the dog. Remember, we get what we reinforce. So if we are using luring and shaping to get a behaviour, we are not only reinforcing the end behaviour (known as the terminal manoeuvre) we are also reinforcing everything that has come before. The cue is connected to all the other 'help' that is being offered.

Let's take the example of teaching a dog to follow left and right verbal cues. We could use the following:

- A verbal cue.
- An arm pointing in the direction we cue.
- A shoulder indicating the same direction.
- A target on the left and right.

All these cues are reinforced when the dog responds by turning left or right. We need to be aware of this, and fade these extra cues, one by one, until the dog responds solely to the verbal cue. In addition, we should bear in mind that we will also reinforce any mistakes made in the learning phase and these will be associated with the new command.

Therefore, we should continue working on the behaviour until the dog meets all the criteria. Then, when he begins to anticipate the behaviour, we can introduce the new signal. If we reinforce this repeatedly and omit the reinforcer if the dog fails to meet the criteria - for example, demonstrating the behaviour without a preceding signal - he will soon learn what he needs to do to get the reward. Now the new signal is important!

PUTTING A NEW BEHAVIOUR ON CUE
Step-by-step

- Condition the new behaviour by luring and shaping, (and body language) without introducing a verbal cue/signal.
- Work to a level where the dog begins to anticipate the new behaviour.
- Now add the new signal. We use classical conditioning for this. Thanks to Pavlov, we know that we need to be very quick and insert the signal within half a second to make the link most effective. Repeat this 3-5 times.
- For the next few trials, don't use the new signal. The dog will try to earn his reinforcer and will show you the newly-learnt behaviour. Now *instead* of rewarding, just restart the dog and wait. He will try again and again until he pauses very briefly.
- Right here, within the space of the pause, reintroduce the new signal. Reinforce as soon as the dog demonstrates the behaviour.
- In the next few training sessions, work with a reinforcement schedule that teaches the dog that he only gets the reward if he is given the signal, and there is no reinforcement if he shows the behaviour without a preceding signal.

Reintroducing the cue at the moment the dog pauses allows us to make use of what always happens when behaviour is no longer reinforced. The dog discovers that the behaviour that had been proving worthwhile is no longer being rewarded. After a few repetitions, he will change his behaviour. The trainer waits for a brief moment of hesitation and, at this very moment, gives the new signal. The dog demonstrates the new behaviour and is rewarded. Now he begins to understand that the new behaviour is worthwhile only after the signal, because that is what leads to the reward. In this phase, we may also see an extinction burst (see *page* 79) where the dog offers all sorts of behaviours, hoping to do something that will earn the reinforcer. If this happens - just ignore it - it will go away on its own.

Once the new behaviour is established and put on a cue, we can teach the dog to discriminate, i.e., we make sure he only responds to this specific signal. We know that dogs are guided by our body language in addition to our verbal cues. Therefore, we now need to change position. For example, we no longer face the dog directly when we give him the cue, but sometimes stand next to him or behind him. We can sit or lie down and then give the cue. We can look in a mirror or hide behind a tree or a car and then give it. This allows for powerful conditioning. No matter what we do or where we are, the dog responds with the desired behaviour when he hears the verbal cue. This is a very important moment in all types of training. We found it to be particularly the case with our radio directional camera dogs because they are working with their handlers out of sight. When we reached a point in training where we were confident that a behaviour was well-conditioned, we tested the dog's ability to discriminate further, giving commands via the radio attached to the dog's collar. In this situation, the dog had to give the correct response even if he heard the voice of his handler through a radio speaker. This 'mechanical' voice sounds totally different then the real warm voice of his handler. The volume, pitch and emphasis of how we say a command is important.

NEW CUE BEFORE KNOWN CUE

As humans, we are capable of understanding many of the details, situations and events around us. If a new situation arises, we can find out the reason for it. We can identify the triggers, put them in the right order, and learn how best to respond the next time we are faced with a similar situation. It's different for dogs. As previously highlighted (see *page 91*) if we want to teach a new cue, it must precede the cue the dog already knows.

To recap: when we put a newly learnt behaviour on a verbal cue, we:

- First give the verbal cue (new word).
- Then the body language signal (old signal) that we have always used.

The same applies if we want to teach the dog not only to respond to a call, but also to a whistle:

- We use a whistle first (new signal).
- Then give the familiar cue, *"here"* (old cue).

We also condition laser or ultrasound as a signal with the same procedure, putting it in front of the known signal. If we tried to turn it around and give the old cue followed by the new cue, it simply would not work. A known command evokes well-conditioned behaviour that usually results in a reward. The information the dog gets

Our camera dogs get their directional commands over the radio attached to their collar – at first from nearby, then from increasingly longer distances until the handler is out of sight.

after the known cue is not relevant and he will not do anything with it.

If we want to introduce an additional or a different cue for a behaviour that is already known, the new signal must always precede the known cue. It doesn't work the other way round!

Unlike dogs, we humans are well able to understand a new cue even if it comes after the one we know. Let's imagine that we are startled by a noise. Someone yells: *"look out!"* and then a branch falls from a tree. Next time we hear someone yell: *"look out!"* we will immediately run for cover, whereas our dogs would be hit by falling branches time and time again. We need to keep this in mind when we find ourselves pulling on the leash and then giving a command. The dog will do nothing with the command, but only react to the leash pull.

THE RESET

Just like humans, animals learn through trial and error. This is most easily understood if we take the example of a baby learning to walk - a process we have all been through. We start by crawling, pull ourselves up on a chair or table, stand for a few seconds, fall, and start again. After many attempts, our vestibular system starts to help us, our balance improves, our muscles get stronger, and we can take our first steps. We fall, crawl, get up, walk, fall, and try again. Eventually we learn to walk - but imagine being penalised for every failed attempt! What if our parents had scolded us every time we fell and had not encouraged us to continue trying? Our learning would have been blocked, or we might have slipped into a state of learned helplessness (see *page 135*). Also consider what would have happened if our parents had kept us from falling? If we had been grabbed every time we seemed to be losing our balance? Or if we had been restrained in a harness to prevent us from moving? We would never have learnt to balance and walk on our own. We would never have developed the necessary muscles and never learnt to co-ordinate our movements.

To learn new skills, we must make mistakes. We must experience the unpleasant consequences of falling as well as the pleasant consequences of being able to move faster than when we crawled, otherwise we would continue to crawl, as this is a skill we have already acquired. In addition, standing up and walking not only allows us to move faster, we can also now see the world from a different perspective.

Trial and error have a part to play in dog training, as long as it is well-directed, planned and thought through. As dogs progress in training - particularly operational dogs - they will be confronted with problems they need to solve using their own initiative. In a working situation, the dog will need to find a solution, deal with the circumstances and continue with the task. I am a great believer in the error-free learning strategy (see *page 123*), but in training we deliberately introduce some elements of trial and error so that dogs learn to persevere and complete the mission.

Punishment has no place in this process; it would block learning, or the dog would learn to avoid the situation. However, the trainer should not interfere to stop the dog making a mistake. In a real-life situation, the trainer would not be on hand to find a solution. The best way forward is to work with the technique of restarting or 'resets' in combination with clever training protocols, which are structured to progress in small, incremental steps. As already highlighted, each protocol covers several training sessions, each consisting of a set number of trials. There must always be sufficient trials to ensure the behaviour is well-conditioned. We cannot move on to the next sub-step after two or three successful trials, because the behaviour may have happened by chance, and we cannot build reliable behaviour on luck. We must therefore repeat the trials until the behaviour is fully established. Then, in the next session, we can increase the criteria, adding more distance, duration, more weight, more distractions, adverse weather conditions, and so on. The dogs learn to 'hang on in there' and to keep trying because they know their efforts will lead to reinforcement.

A window of opportunity opens up at the beginning of a trial: the dog has the chance of getting a reward. If we build this up in small steps, and the dog understands the task, the rules are clear. We then increase the challenge and ask him to work a little bit harder; if he is unsuccessful, there is no reward. The window of opportunity closes and the dog is reset for another attempt.

THREE TIMES AND YOU'RE OUT!

If a dog repeatedly makes a mistake, there is no point in restarting again and again. It may be that the progression is too much, the distractions too strong, or the conditions are complicating the task. There could be an error in the training protocol, or the individual dog may need additional sessions to develop the skill and confidence to take this step.

To keep things simple, we work on a rule: *three times and you're out!* In practice, it means that if we have had three successive resets within a block of 10 trials, we stop. We take a break and put the dog in his crate or in the car. We sit down and look at our data. We analyse the situation and make changes in the next set of trials, i.e., we make the criteria easier so the dog can be successful again.

BEHAVIOUR ECONOMICS

Behaviour economics plays a major role in all my training plans. If animals work, burn energy and get nothing in return, they will change their behaviour. Let's look at the example of a cheetah stalking a herd of antelope. He has his eye on a calf; it's still a long way off but the cheetah is hungry. He cannot hold back any longer and starts running. Within three or four seconds he has reached a speed of 90-100 km/h. The young antelope has seen the predator and flees. The attack distance was too great. The cheetah got close, but it is a sprinter, not a long-distance runner, and it could not maintain the high-speed chase. The antelope calf survived to live another day. The cheetah learnt to be more patient the next time it hunts; it will stalk until it gets closer to the herd or seek out an older or sick antelope that cannot run as fast, or as far.

The principle of behaviour economics applies equally to dog training. When I am training scent detection dogs, I prepare a line-up which includes the target scent. If a dog works too quickly, especially at the start of the line-up, I let it happen. I observe as the dog runs past the target scent and let him run to the end of the line-up. He is allowed to search back and forth, and, after four attempts, I halt proceedings. This closes the window of opportunity, and the dog has lost the chance of getting a reward. Following a reset, the dog will work more slowly and with more focus. If this happens, and the dog finds the target scent within four attempts, the new criteria is: *at most three (or two) trials for a reset*. Nothing found? Reset. The dog therefore learns to work more slowly without introducing a leash or using any other means to slow him down. The goal has been achieved using the power of behaviour economics!

At the beginning of a line-up, dogs tend to rush and overrun the target scent. To make them slower, we employ behavioural economics.

PROTOCOL COMMUNICATION

Much has already been said about the importance of clear communication, and we need to ensure this applies to protocols, regarding both dogs and handlers. After a cue has been given, the outcome is straightforward: if the dog has not met the criteria, there is a reset. Once he meets the criteria, there are four ways we can ensure he never gets bored, because if we gave a positive reinforcer every time the dog met the criteria, we could not continue to chain behavioural elements and our training would become very predictable.

Imagine a dog receiving a verbal cue to sit. He goes directly into the 'sit' position and is reinforced with a treat. However, after some repetitions, the dog starts to lose interest or becomes sloppy in in his response. To counteract this, we need to keep him guessing:

- When is the reinforcer coming?
- What will the reinforcer be?
- Where is the reinforcer coming from?

In this way you can reinforce those 'fast' or 'special' responses and, by using differential reinforcements such as different sorts of food, a ball, a kong, some petting, a chase game, a tug of war game, or whatever is reinforcing for that particular dog, we can make a difference and keep the dog engaged in training!

Following the correct response to a cue – in this case, "sit" – and the criteria have been met, there will be four options to continue (green), not just one. This keeps training interesting!

THE RELEASE CUE

There are different opinions regarding the use of release cues. I use it as a cue to the dog to:

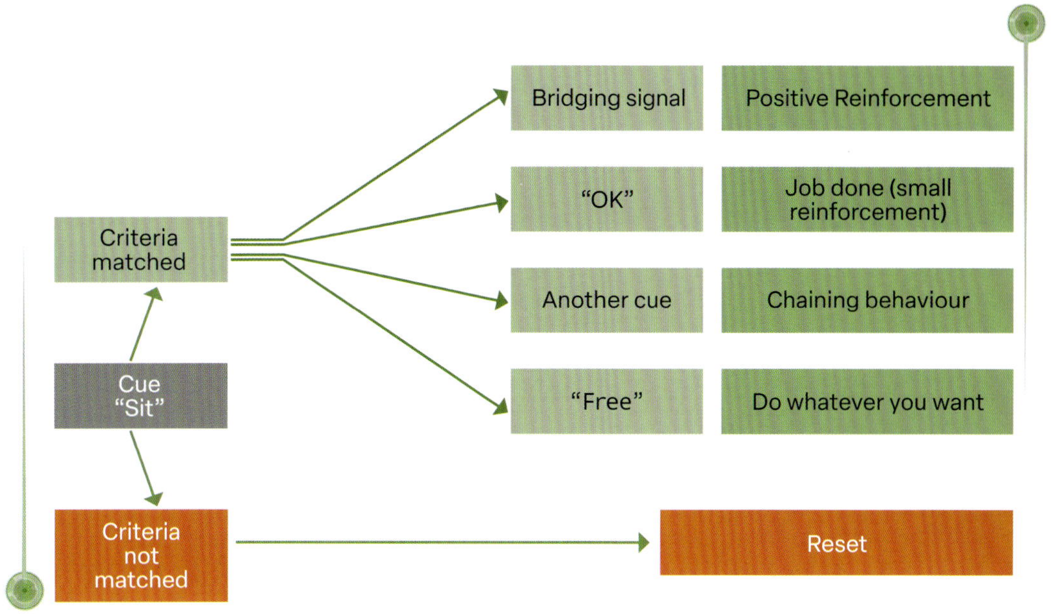

Following a correct response to the verbal cue ("sit"), and assuming the dog has fulfilled the necessary criteria, there are four possible outcomes. This keeps the dog guessing and makes training fun!

- Start work.
- Go out.
- Grab a reinforcer.
- Leave the transport crate.
- End a behaviour.

It is a communication tied to a specific moment in time and context. I use release cues to establish communication, to set up rules, to practise patience and to have fun. I often work with several dogs at the same time; one has a task while the others wait on platforms. This is 'work' for everyone. When I am ready, I release each dog by calling their name. This is the cue to leave the platform and start work. I use the same release cue (dog's name) when I open the transport crate in the car, or the front door. When a dog hears his name, he is allowed to jump out of the car or go out of the door. This establishes a certain order, or rather structure, in the group. One of my good friends, who works with several hunting dogs at the same time, also uses their names as a call to start work: while five dogs watch the dummy-throwing machine in action, only one is released to run and fetch the dummy.

I also use a release cue for self-control. This might be prior to feeding, leaving the house, jumping out of the car, as well as to end a 'sit' or a 'down' after a dog has waited to cross a road. A dog will understand from the context whether we are calling him by name to get his attention, or if his name is serving as a release cue. Besides using names, I also use the verbal cues 'free' and 'ok'.

WORKING WITH A TRAINING PROTOCOL

In my protocols there is always a column allocated to training time. I believe we need to find a rhythm in training and schedule repetitions. This will help us to find a flow so we can reach fluency in training and wanted behaviour. Scheduling training time gives us a better understanding of how quickly or slowly we can complete the steps in the protocol. We should also allocate different amounts of time for training new and priority behaviours, and behaviours that have already been conditioned but need to be maintained at the required standard and accessed reliably.

Novice trainers, who are new to behavioural chains, tend to focus on one behaviour at a time. However, each element needs to be balanced so that all links in the behavioural chain remain strong, which requires planning and preparation. Of course, this is time consuming at the beginning, but overall, it does save a lot of time in the training process. Time is precious, and good planning allows us to be more efficient and effective.

All of this brings us to the question: what is 'good enough'? The answer to this question must be found in the training plan:

- What can we expect after the training ends?
- What is the skill level of both dog and handler?

MOTIVATION IS KEY

The foundation of successful training is focus and perseverance, and both come from motivation. Without motivation, it becomes very difficult to meet the criteria. As mentioned above, if the dog no longer meets the criteria, we have to look at the reinforcers. Often they are not of sufficient value or the rate of reinforcement is too low and, as a result, the dog loses motivation. Remember the principle of '*what's in it for me?*' (see *page 135*). We, as trainers, must make sure that what we ask of our animals is always interesting and fun.

We can influence motivation using two techniques: positive reinforcement (R+), but also negative reinforcement (R-). School grades are a good example of this. If a pupil studies well and does their homework, they will get better grades – positive reinforcement. If they don't want to study and fail to do their homework, bad grades can be the negative reinforcement that motivates them to do better. Both reinforcement techniques are externally motivated: motivation is increased by getting something good or by making something bad go away. But there is also an inner motivational booster known as 'intrinsic motivation' (see *page 90*). For operational work I favour dogs with high intrinsic motivation for the required tasks and I use positive reinforcement techniques to change behaviour.

I always set up training protocols in such a way that it is easy for the dog to succeed, factoring in a high rate of reinforcement, especially in the initial stages of working on a new behaviour. I also check the reinforcers to make sure they are valuable to the dog. If not, I make changes – using cheese instead of kibble, for example – whatever motivates the dog to put in more effort. However, the key element to ensure success is to write smart training plans, which make use of small, progressive steps.

- When is it good enough and who decides?
- Who sets the bar for training goals?

For information on testing and evaluation, see Chapter 21.

EXAMPLES OF TRAINING PROTOCOLS

I strongly encourage trainers to design their own protocols and to write them down. There is a lot to be learnt from this process and it has the advantage of being personalised, working with the strengths and weaknesses of a particular dog. Protocols can also be shared with friends, which will encourage a shared a learning experience, gaining from other people's expertise.

To provide a better idea of possible protocols, here are a few examples from my own work.

SIT PROTOCOL

Training a 'sit' behaviour can be very easy, but also very challenging. The behaviour itself is simple, but the challenge is to condition it under all possible circumstances and distractions, and with extended duration.

Let's start with the criteria:

- What does a 'sit' look like? Bottom on the floor or half to the side with just one hip on the floor? I like a neat sit, with front legs straight, and focus on the handler.
- The dog should sit still, not bark and hold his tail steady. His mouth should be half open and relaxed.
- Duration: Do we want the dog to hold position for 10 seconds or two minutes?
- Response time: We want a quick sit within two seconds of giving the signal.
- We want the 'sit' to be generalised so the dog will sit outside on concrete, asphalt, grass or sand.
- The dog should remain seated regardless of distractions.

So, when considering the criteria, the simple 'sit' behaviour has become far more complex, and we can see the need for thought, planning and preparation prior to writing a training protocol.

Whenever I ask participants in a workshop whose dog can do a good 'sit', many hands go up. But then when I ask them to get their dogs and show what they can do, things invariably go awry. Suddenly, I am listening to a string of excuses and only a handful dare to demonstrate the 'sit' with their dogs. Then I get a chance to see exactly what is going on. Some dogs need at least three signals before they sit down. Others only sit for two to four seconds and then do something else. Some only sit in response to verbal cues in combination with a small tug on the leash. Some don't mind other dogs or people around, others do. Training a good, reliable 'sit' takes time and a protocol. This is the only way we can set clear goals and check progress.

To give an idea of what a 'sit' protocol might look like, here is an example from my seventh session which consists of 10 trials. It helps to visualise the situation, so I start with a written description and, ideally, a photo. I also work with video to explain the different sessions and trials. In this session, I would explain that the trials need to be completed twice, with a minimum success rate of 80 per cent before moving on to session number 8. In this example, the dog attained a 60 per cent success ratio which means I need to keep working on session 7 until the behaviour is strong enough for us to progress.

SIT

Protocol Session 7

Situation:	Children's playground.
Behaviour:	Sit with focus on trainer.
Cue:	Verbal cue – "sit" – from handler, who then walks away backwards.
Distraction:	Tennis ball.
Place to sit:	Platform.
Leash:	Short leash.
Distance from trainer:	5m.
Bridge signal:	Clicker.
Reinforcer:	Food (kibble).
Duration:	Trainer remains stationary at distance for 2-10 seconds.
Trials:	10
Reset:	If the dog leaves the sitting position.

TARGET PROTOCOL

Now let's take a closer look at one of my target training protocols. I work with different targets; in this example, the dog must press his nose into the target when signalled.

PUSH NOSE INTO TARGET

Protocol Session 3

Situation:	Outside, fenced training ground.
Behaviour:	Sit and push nose into the target.
Position of dog:	Sitting in front of the target.
Trainer – Action 1:	Place the dog in a 'sit' and stand behind him.
Cue:	None – the dog anticipates based on target/context.
Distraction:	None.
Place to sit:	Hard surface in front of the target.
Leash:	None.
Distance of trainer:	1m behind dog.
Bridge signal:	Sound of the feeder.
Reinforcer:	Food from feeder.
Trainer – Action 2:	After the automatic feeder has been activated, bring the dog back in front of the target.
Duration:	None.
Trials:	10.
Reset:	If the dog breaks the 'sit' and walks away.

Again, we collect data:

- Did we complete 10 trials?
- How often was the dog distracted?
- Did he need a reset?
- How quickly did the dog anticipate, i.e., stick his nose into the target, after the trainer had positioned him in front of it?

In later training sessions, we will still cue the behaviour but change the criteria – for example, wait for the cue, keep the nose in the target longer, add distractions, increase distance from trainer.

DESIGNING A PROTOCOL

Using laser as a target

I started using lasers in 1996. This opened up some very interesting possibilities, not only tactically, but also because I could cover longer distances and work with major changes of directions. Adding a high-quality reinforcer to the laser created a powerful target. So, it became easy to send my dogs to a certain place or use the laser to give them information about which direction to take, or what to do next. A laser can be directed at a wall or an object, quickly changing distance or circumstances. Training the laser as a target is a challenge, but one of the biggest advantages is that we can very easily combine the set-up with behavioural economics (see *page 215*) – the laser dot itself becomes the target.

The laser turns the previously-trained wall target into a 'hot' target.

I have worked with many different methods to condition the dogs to a laser point using both classical and operant conditioning: food, balls, electronic devices and many more. The most interesting method for trainers is to work from a clear protocol based on operant conditioning. At this point I will not provide the exact training plan – it is your turn to write your own! However, I will give you a few tips:

- I first train a wall target. The behaviour I want is a nose touch. When this is established, i.e., the dog touches the target with his nose and receives a bridge signal and reinforcer, I start the next phase.
- Now the laser comes into play. The laser point turns the wall target into a 'hot' target, which means that a reinforcer is given when the wall target is illuminated by the laser point.

Developing the behaviour that allows the dog to be guided with the laser point is a challenge for the trainer, involving a range of practical skills such as operating the laser, selecting the right target, operating the bridge signal and giving the dog the reinforcement, as well as making quick decisions based on observational skills.

Here we can bring some of the previous lessons to build a good laser training protocol:

- The new **antecedent** (laser) triggers the dog to show the new **behaviour** (touch target when hot) and a nice reinforcer follows as a positive **consequence**.
- Then we bring the principle of behavioural economics into play: when the dog is at the starting position, we

train him to start immediately. As soon as he looks at the target, we switch on the laser pointer and illuminate the target. If the dog does not react within two seconds, the laser is switched off. So, if the dog is too slow, he will not get a reinforcer; laser off means the window of opportunity is closed, the chance of earning a reinforcer has gone.

Now we can add some magic to the protocol. If we increase the distance between the start and the target, we can change the dog's speed:

- First, we measure the distance between start and target, let's call it 25m.
- Then, when we turn on the laser, we measure the time it takes the dog to touch the laser dot. Let's say 15 seconds.
- We repeat this: in the second run the dog needs 18 seconds and in the third run he does it in 12 seconds. Therefore, the average time it takes the dog to reach the illuminated target is 15 seconds.
- Now we can put a timer on the laser so it will switch off after 14 seconds. If we gradually reduce this time, we can make the dog faster and faster!

DATA, DATA AND MORE DATA!
Collecting and analysing data is the only objective way to assess how training is progressing because it provides reliable information. It is also an indicator of whether the dog and handler are ready for the test, the mission, or whatever the goal may be. There are many ways to collect data, be it with the help of a smartphone app or with an old notepad.

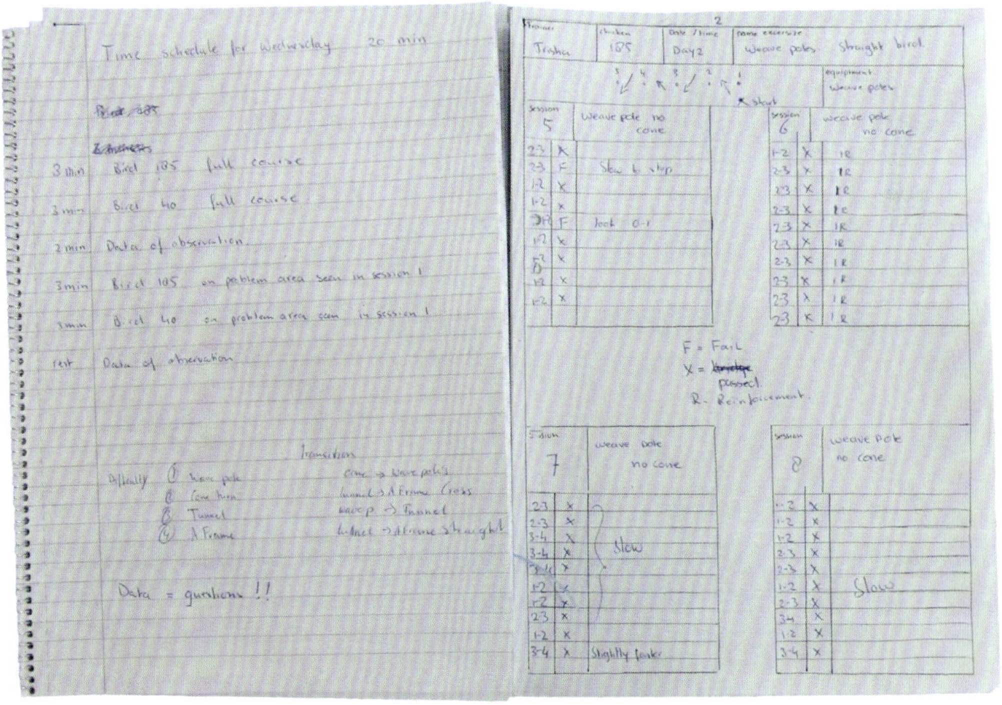

It doesn't matter how we collect data – what matters is that we do it!

INTERPRETING TRAINING SHORTHAND

I use letters and symbols in my data logs to speed up my administration. However, it is important that we use them correctly to ensure they can be clearly understood. Once we have started using them, we must stick to them. Here is a list of the 'shorthand' I use in my protocols:

S	stimulus (dog must be able to see, feel, hear, smell or taste it).
R	response (selected behavioural unit).
Sd	discriminative stimulus (a stimulus paired with a primary reinforcer).
Sn	neutral stimulus (no meaning for the dog).
Sr+	conditioned reinforcer (now acts like primary reinforcer).
R+	positive reinforcer (gives us more and stronger behaviour).
R-	negative reinforcer (gives us more and stronger behaviour).
P+	positive punishment (gives us less and weaker behaviour).
P-	negative punishment (gives us less and weaker behaviour).

In the datasheets for direction training, I also use:

L	'left'
R	'right'
F	'forward'
H	'here'
T	'touch'
S	'sit'
D	'down'
A	'apport'(fetch)
P	place an object

Chapter 21: TESTING AND EVALUATION

In general, tests and exams mean stress, anxiety, arguments and all sorts of excuses and evasions. People shy away from being tested – I have come across only a handful of trainers who had the confidence to volunteer for a double-blind test. What is the reason for this? Perhaps both the dog and the handler are not properly trained, or the stress that the handler is experiencing has a negative impact on the dog? Or is it because errors are triggered by the exam situation, which may differ from the training process?

Despite these misgivings, I believe we should embrace the opportunity to be tested. We should check ourselves, our dogs and our training concept on a regular basis – and we should enjoy doing so. We must learn to trust our dogs and believe in ourselves. We must face the challenge, because there will be a day when we will be tested in a real-life scenario, which could be on operational duty, in a competition, or coping with an everyday issue.

Of course, anyone who starts a new job and learns new skills will initially question themselves. Am I ready for the (operational) tasks assigned to me? Can I take part in this competition? Am I sure of my ability? If a dog is to be the other half of the partnership, there will be even more questions and doubts because we now have to worry about the dog's performance. The only way forward is to trust one another; nobody can help with this, and there is no one to blame if it does not happen. But with trust comes confidence. I have learnt to like exams; they give me an opportunity to show what I have achieved with my dog, and to discover our strengths and weaknesses. It is an opportunity to grow!

For those who are unconvinced, read the following excerpt from President Theodore Roosevelt's *Themes in the Arena* speech, delivered at the Sorbonne in Paris on April 23, 1910:

"It is not the critic who counts; not the man who points out how the strong man stumbles, or where the doer of deeds could have done them better. The credit belongs to the man who is actually in the arena, whose face is marred by dust and sweat and blood; who strives valiantly; who errs, who comes short again and again, because there is no effort without error and shortcoming; but who does actually strive to do the deeds; who knows great enthusiasms, the great devotions; who spends himself in a worthy cause; who at the best knows in the end the triumph of high achievement, and who at the worst, if he fails, at least fails while daring greatly, so that his place shall never be with those cold and timid souls who neither know victory nor defeat."

When faced with criticism, we need to remember that it is the team – the dog and handler – out there doing the work, daring greatly, so we need to be proud of what we do and see it as a preparation for the work that will come later.

There is a significant difference between classroom testing, sports testing and professional service dog testing. In the classroom and in the sports arena, circumstances can be controlled to a large extent, whereas operations conducted by the police, customs and military are varied and unpredictable. For example, there will be many distractions which are outside our control, including those we may not even notice because, unlike our dogs, we may not see, hear, smell, or feel them. In addition, failing a test in the professional arena has far greater consequences as status, ego and position may be at stake. If the head trainer, who is responsible for the entire service dog unit, fails the exam, how will that affect his/her students? If the explosives dog-handler team fail to find the hidden bomb in a test scenario, how will it affect the other members of the unit? If the narcotics dog does not find the hidden drugs during a prison search, will the prison governor call in the dog team again? If the tracking dog cannot find the suspect, will the next investigation even ask for tracking dogs? It is important that dog-handler teams are ready for their tasks, and that the outcomes are predictable and reliable.

The most effective way to achieve this is regular testing in the form of odour recognition tests (ORT) and operational validation tests (OVT).

For those dog handlers who are resistant to taking these tests I ask the following questions:

- *"Would you keep working with a piece of equipment that you cannot check, or that is unreliable?"*
- *"Would you trust a gun that failed to fire 8 out of 10 times? Would this be the primary firearm you carried on operations?"*
- *"What about that unreliable car? Would you take a gamble on it starting, or would you get a new one?"*
- *"What about a mobile phone that keeps losing connection? Would you keep using it or would you buy a new one?"*

Obviously, in all these situations, we would look for an alternative, something we can test to ensure it is reliable. We should, therefore, expect dog-handler teams to be willing to be tested and show their skills so that everyone involved, from the handler through to the top of the chain of command, knows what they are capable of in real-life situations.

The test for odour recognition.

Testing also helps to eliminate assumptions. A patient in hospital would describe their symptoms, pinpointing the trouble-spots and dismissing things that are not causing concern. The patient may even attempt to diagnose the problem. The doctor would listen but would then go on to collect objective data before taking the next steps. In the same way, reviewing dog-handler teams gives us an objective view of what is going on, and what standard has been reached.

Handlers often tell me that their dog's performance dips in operational situations. That tells me that there is a discrepancy between training and operations, and the interface is not reliable. However, we should bear in mind:

- A radar device works just as well in training as it does in an operation.
- A weapon works just as well in training as it does in an operation.
- A car starts equally well in practice and in an operation.
- A telephone works in both situations.

If an 'aid' is proving unreliable, we need to review what is going on and make decisions. If there is a difference between training and field work - and the team cannot perform reliably in a double-blind test - then we have made a mistake in our training plan. Everything must be geared towards developing a reliable dog and a trustworthy handler who can serve as a reliable interface within the chain of command. In order to achieve this, we have to take both dog and handler out of their comfort zone more often during training. This will prepare them for the real operation, the exam, the contest.

If there is a difference between training and operations, we have made a mistake in our training plan.

Consequences change behaviour; this not only applies to animal training but also to the way we humans respond. If dog-handler teams are not operational because they keep failing tests, there must be consequences. This could mean that team is not operational or, alternatively, we could change the training plan. Reviews must reflect the criteria demanded in operational situations. They must be realistic and integrated into everyday training so that they become second nature to both dog and handler. This also means that we must redefine training so as to maintain learnt skills, ensuring that dog-handler teams are consistent and reliable in their work.

If a police officer fails to achieve the necessary hit rate in the shooting range, he/she is not allowed on armed patrols, an unreliable driver is not allowed to drive with blue lights and sirens, a failed fitness test means a temporary suspension from operational duties. The consequences of a drop in performance are clear cut. However, this is not the case with an unreliable dog and handler team. Handlers feel offended and blame conditions, weather, distractions, or question the examiner's competence. To overcome this issue, we need an objective assessment, such as working a double-blind which gives definitive proof as to whether the team is fit to be operational.

FORMAL TESTING

Professional organisations, such as the police, the military, customs, have developed their own tests; this is also true of teams who train medical detection and species protection dogs. In my experience, these tests cannot cross international boundaries as the parameters not only change from country to country, but also from region to region. For example, there is a big difference between training a tracking dog and his handler to work in an urban area compared with a rural area. A drug detection dog who works in an international airport has a different job compared with the dog who is trained to work in a large seaport with thousands of containers and moving trucks.

Of course, we need to set a standard based on what is good enough. For dogs who are trained for operational duties, this consists of two tests:

LEARNING FROM MISTAKES

Everybody makes mistakes. More important is what we do *after* a mistake. Mistakes help us learn, grow, and progress.

I remember being really disappointed when one of my dog-handler teams failed to find the target scent I had hidden in our offices the previous evening. I questioned the handler's training efforts and his overall work ethic. Following this uncomfortable conversation, the security manager knocked on my door and proceeded to hand over a number of items. These were the objects, infused with target scent, that the dog-handler team had failed to find. I then discovered that, as a joke, one of the security guards had confiscated these items after watching footage of me on the surveillance camera. The joke backfired, and I had made a terrible misjudgment due to a lack of information. I was quick to apologise to the handler, and I made sure I learnt from my mistake.

How can we distinguish between unintentional mistakes and intentional misdirection? Deliberate misdirection is when someone is caught doing something undesirable and then apologises for their 'mistake'. For example, a handler punishes their dog out of frustration, gets caught and apologises. In this instance, the handler is not taking responsibility for their actions and is not accountable for what has happened. Taking responsibility entails owning up, learning from the mistake, and working out what needs to be done to ensure it does not happen again.

Sometimes we see intentional misdirection in an operational situation. I was involved in a high-risk operation where the task force had to storm a house while the suspect was sleeping, and then arrest him. The plan worked and the suspect did not have time to reach for his gun. We then asked a specialist dog handler to conduct a search. While searching, the dog came across a large bag in the corner of a room. He examined it, but did not react, and continued working. The handler kicked the bag, and we heard the characteristic metallic sound of ammunition. We opened he bag and found over 3,000 bullets. But before returning to HQ for the debrief, the handler told everyone that his dog had made a very good passive alert which led to the recovery of the ammunition. Instead of taking responsibility for the fact that the dog had missed his target, he deliberately misled us. It was a shame that the handler felt the need to 'cover up' for his dog in this way because we can learn from our dogs' unintentional mistakes – but not when they are intentionally swept under the carpet!

We can learn from mistakes, but only if we don't sweep them under the carpet.

Odour Recognition Test (ORT): This consists of working a double-blind. Each target scent is tested in six trials, and one of these trials will have no target scent.
Operational Validation Test (OVT): This is carried out in the context of the specific operational task required. This test is also double-blind and will look like an average operational task for the dog team, extracted from data from the operational work environment. For example, a narcotics dog that is to be used in car searches will do his OVT on cars and not in a house. A explosives dog who will work in government buildings will do his OVT in office buildings. A tracking team scheduled to work in a big city environment, requiring a lot of hard surface tracking, will not do their OVT in a forest or a rural area.

The training protocol, and the OVT, should be designed so that whoever is responsible for the working remit - the commander, manager or project leader - knows what to expect from the dog, in consultation with the handler. For example, if a military convoy is stopped in a war zone because of a dog's alert, the handler needs to be confident regarding the nature of what the dog has found, and the likelihood of it being a genuine alert. The unit commander will not be impressed if the handler reports: *"My dog is interested, sir, but it is not an alert"*.

The same applies to the airport security officer who stops a flight departing because their dog has alerted to a piece of luggage. In this case, if the handler reports that the dog is 'interested', it does not provide sufficient evidence to make a decision that has far-reaching consequences. Therefore, the handler must be the interface between the dog and the command level, providing a clear picture of what is going on rather than a vague supposition. With this in mind, we need to ensure that the training plan clarifies the value of the alert, i.e., what is the difference between 'interest' and an alert, and how can we evaluate this information so that it can serve as a basis for decision-making?

TESTING TRACKING DOGS

Working a double-blind is the obvious way to test a tracking dog and his handler. The protocol must include a clear description of what the head of operations, manager or project manager expects from the tracking dog and his handler, which must be in line with the real operational tasks in that area. We, therefore, need to collect and analyse data from previous operations. When I did this, here in the Netherlands, I found that that after a criminal offence, it took the police 15-30 minutes to complete their initial investigation, and they were followed by the forensic team. On average, the decision to request a tracking dog was usually made 45-60 minutes after the incident had been reported to the police dispatch. The quickest respond and travel time for a police K9 team was around 60 minutes. This meant that the track left by the suspect was two hours old before the first tracking dog was deployed to the crime scene. In addition, the data showed that eight out of ten deployments took place in an urban environment, so we needed to prioritise testing our response teams on tracks that had been laid for two hours in an urban environment. To make sure that they could do a first-class job, we also needed to challenge them with a track that has been laid for three hours to make sure they could do what the detectives expected of them.

THE DOUBLE-BLIND TEST

The runner (the track-layer) tells the dog-handler team where to find the starting point, such as the 'suspect's' car or an abandoned hat, bag or gun. We don't need anyone along the way to check that the team is staying on-track because nobody knows exactly how it behaves under the constantly-changing temperature and wind conditions. In fact, we have learnt from experience that dogs and handlers are influenced by the third parties who may, unconsciously, provide information – remember the clever Hans effect (see *page 131*). The aim is to replicate an operational situation as closely as possible. The training plan must include the length of the track, whether items should be found along the way, and if so, how many. Nowadays we can also work with GPS devices to collect even more information.

Tracking teams have to be tested under conditions that are as close as possible to real operations.

TESTING SCENT DETECTION DOGS

As with tracking, we humans like to think we know a lot about how scent behaves outside in the environment or inside a building. We think we know how and where it flows out of a bag, a suitcase or a cupboard. But the truth is we don't know for sure; a lot of our suppositions are subjective, which means we may not make an accurate assessment of what is going on when a dog is working. Therefore, our scent detection dogs are tested with the odour recognition test (ORT) and, if successful, this is followed by the operational validation test (OVT) which takes place in an environment that mimics real-life situations as closely as possible.

There are many ways to test a dog's olfactory ability and I have seen a variety of methods that have been utilised across the globe: short or long, very thorough or very quick, just a few target scents or lots of them. In some tests there was leeway for mistakes, in others a 100 per cent success rate was expected. All in all, there was a lot of variation - and a high degree of subjectivity. However, such tests can help us learn about the quality and capabilities of our scent detection dogs, which will ultimately make the world a safer place. In addition - and not to be under-estimated - the data obtained from these tests is a major resource. If we could collect all the data that has been obtained, we could learn a lot from each other.

Problems with K9 validation exist worldwide, which makes testing all the more necessary. In New South Wales, Australia, a report on drug detection dogs was published in 2006 stating that drugs were found in only 26 per cent of a dog's alerts. Of these, 84 per cent of the cases involved very small amounts of cannabis and only 0.19 per cent of the finds resulting from dog alerts led to criminal prosecution. In 2010, the University of California tested 18 drug and explosives detection dogs and reported an 85 per cent failure rate. Unfortunately, there are many other similar examples. We know dogs are amazing at helping us track down criminals, find drugs, explosives, pathogens, etc., but we also need to make sure the human-canine teams are well-trained and ready for their important work.

So, what is urgently needed is a validated test for scent detection dog work, which is carried out according to a clear, objective examination protocol and provides information on the status of training and readiness for operations. It must provide measurable feedback on the training and quality of the dog and handler and, most importantly, their reliability, which in turn would increase overall trust in detection dog teams. This type of testing could extend beyond the annual exam and be used more generally to check performance following a few weeks' vacation, for example, or to measure conditioning of new target scents. There are times when we need to recruit external scent detection teams for an operation and, in this instance, we can test them beforehand which gives everyone involved a sense of security and confidence.

Odour recognition test (ORT)

In principle, the test for odour recognition can be the same for all scent detection dogs – we don't need different testing procedures for explosives, drugs, cash, guns, accelerant, cadaver odour, blood or whatever the dog is tasked with finding. Of course, all these odours have their own characteristics: some have heavier molecules and others are more volatile. Some require dogs to get their noses really close to the source to locate them, others can be easily identified from a distance. The scent picture is influenced by time, UV radiation of the sun, temperature, humidity, oxygen and bacteria. In addition, there are distractions from other smells, movement, noise, as well as working in stressful conditions. The location for the test must, therefore, be a controlled environment, selected with care so that the review can be as objective as possible. The dog is tasked to work on a line up or a scent wheel, which contain distraction odours and a target scent. This is not an operational test. It is a skill test to see if the dog is able to find the target odour without getting distracted by other odours. After observing a series of tests, we can fill our database and get a picture of the important details and parameters. This gives us a better idea of the level and quality of the team we are testing, which is the means of improving our work overall.

A good protocol of the test procedure needs detailed information which will facilitate an objective assessment. This includes:

- The distance between target scents.
- How to deal with target scents.
- Which distractions will be included.
- Whether bystanders will be present.
- How to deal with errors and false alerts.
- How the cleaning procedure works.
- How the scores are interpreted.

Line-up

A line-up provides a good test for both dog and handler as it allows for so many variations. It consists of a row of eight targets in the form of containers, ideally 1m apart. There is a scent in each container, seven of which are distraction odours, and only one is the target scent. Possible tests include:

- The handler works the dog on or off-leash.
- The handler is stationed at the start position while the dog searches.
- The line-up includes different items, such as shoes, bags, boxes, etc.,
- The line-up is located inside, where we have complete control of the environment, or outside, which will increase the level of distraction.

After just a few repetitions, a line-up becomes a clear stimulus for the dog; he knows exactly what it has to do and how he can earn reinforcements. Once we get to this standard, we can work with extreme distractions and with different concentrations of scent.

When we use a line-up in an odour discrimination test, we need to ensure that the visuals, spacing, set-up and environment are always the same because we want to get objective results and collect data. The only thing that differs in line-ups for testing purposes is the target odour. Typically, we test a specific target odour over six trials. So if we are working with five different narcotic odours we will test each odour six times. In this case we need 30 trials (five target odours x six trials). If the test is well run, each odour can be tested in six to seven minutes.

The line-up itself quickly becomes a clear stimulus picture to the dog.

In each of the trials, the target odour is placed in a different location within the line-up. The distractions odours also change with each trial, i.e., their position changes or the distraction odours are taken out and replaced with others. This means the dog really has to use his nose to find the target odour because the other variable parameters do not help him get to his reinforcement faster. We need to bear in mind that dogs are opportunistic animals and always will try to find shortcuts. So, if a dog can use a cue, other than his nose, that will provide better, faster, or easier information; he will obey the law of behaviour economics and choose the option that expends less energy. It is, therefore, our job, as trainers, to teach the dog that the only reliable sense to employ is his nose. Eyes, ears, and touch cannot be trusted – it is only his nose that will lead to reinforcement!

One of the six trials will be a blank trial, i.e., there is no target scent at all. This simulates an operational situation when there may not be target scent, and the dog has to learn to deal with not finding anything. Therefore, the dog has five opportunities to get reinforcement in the six trials. And of course, the entire test must be double-blind!

The blank trial is an important element of training because it's the only way we can see if the dog is giving a false positive alert in a bid to get his reinforcer, or if he is able to leave the line-up, in the absence of the target scent, without alerting.

Scent wheel

A scent wheel also works well for scent discrimination tests. There are several different types on the market, or it could be a DIY project. Purpose-built scent wheels are made of metal or other easy-to-clean materials and are mounted on a height-adjustable stand. As with the line-up, the scent wheel is set up with distraction odours plus the target scent; the difference is in the layout. Instead of a row of scents, the containers are arranged in a circle. The scent wheel is easy to turn and, because there is no beginning and no end, the dogs are extra motivated to find the target scent. In contrast, the linear arrangement of a line-up can be problematic for some dogs because there is always a visual beginning and a visual end. In training we often see dogs skipping the first position in the line-up and then, if they cannot find a target scent, they just sit or lie down at random at the last position.

When we use a scent wheel in an odour discrimination test, we need a clear protocol. In addition to the information required for a line up, it also needs to include how long the dog is allowed to search, how the scent wheel is cleaned between trials, as well as noting the changing position of the target scent and the distraction odours for each trial.

You can build your own scent wheel with relatively simple equipment.

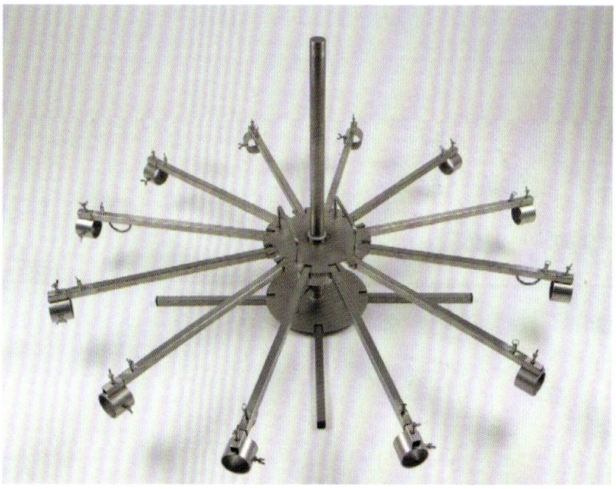

A purpose-built metal scent wheel is easier to clean from odour contaminations. It can be adjusted in height and folded up for transport.

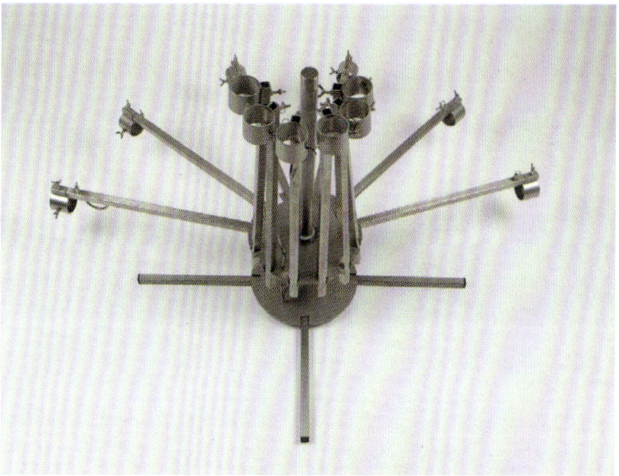

Our newly developed scent wheel, with integrated sensors, measures exactly how long the dog's nose has touched the containers and transfers the data to a smartphone app.

Odour validation test (OVT)

The OVT is a prerequisite for being able to complete the operational test. So, logically, an arson detection dog is tested in a building that was previously on fire, a search and rescue dog is tested on rubble, etc. In the case of drug or explosives detection dogs, we try to use the environment where the dog will later be deployed. This could mean searching through containers in a seaport, or among suitcases and crates in an airport. If the dog is to work for customs, the search is focused on different cars, while an explosives detection dog is tested against the sound of gunfire.

Today, in many tests, I see examiners following the team and checking their work, awarding points for handling, route, distance, odour detection, alert, etc. The problem is that these examiners can, unconsciously, influence the team. It could be in the way they position themselves and move, when they speak and when they remain silent. This applies not only to the handler, but the dog will also use their 'hints' to find the target scent.

I advocate following a different testing principle which is a better way of evaluating the team's work and improves their overall experience. This involves removing the elements of negative stress and replacing it with positive stress. For many people, the word 'stress' is associated with negativity. Negative stress is the feeling of being overwhelmed, anxious, or unable to cope with the demands of a situation. It can lead to burnout, depression, and physical health problems. Positive stress, on the other hand, is the feeling of excitement and challenge that comes from engaging in difficult tasks. It can help us to stay motivated and focused and can even lead to improved performance – something that applies equally to both people and dogs.

By changing the way we mark, and how we reinforce finds, we can build different expectations. Double-blind tests allow us to be more objective and, in addition, we can make them as interesting as possible so that the teams enjoy the challenge and are keen to do them more often than the obligatory once a year exam.

In our test for scent detection dogs, the teams meet the examination committee for a briefing. First, the handler has to present the result of a successful ORT. He/she then receives information about the search area, which could be an area delimited by GPS coordinates, a building or an industrial wasteland. It could also be an airport, a seaport, an aeroplane, a ship or whatever matches the environment of future deployment. In all scenarios, we provide distractions that are as similar as possible to those during an operation. Several finds are hidden in the search area. In the event of a find, the handler cannot simply give a bridge signal and reinforce. Instead, he/she has to examine the item to find an identification code attached to it. This could be a number, a QR code, or a token that must be handed to the inspectors.

In this way we can work double-blind, without being influenced by others and can, therefore, achieve an objective result. Examiners no longer need to make subjective judgments, and the process of reinforcement is more realistic. As we know, there is more to the reinforcement process than giving the dog a bridge signal and a reinforcer. When working with a new dog, or focusing on a difficult find, the handler is highly motivated and gives thought to the reinforcement procedure, carrying it out with energy and enthusiasm. However, after hundreds of finds, the handler may become lazy, throwing a reinforcer to the dog and not fully engaging with him. I like to build procedures so that the handler is always committed to the reinforcement game and 'celebrates' as though it is their first find. So, in our test, the handler must find the code or token, and the dog is the only one who can help. In this way, they are working as a team, which is the perfect way to set them up for real-life operations. In addition, the quality and strength of the reinforcement process are positively influenced. Dogs and handlers love this, and the procedures are real confidence builders! I believe that dogs and handlers can never be 100 per cent perfect and they will always make mistakes. Even lab technicians, with all their expensive equipment, make mistakes. So, we have a choice. We can make the exams so easy that full marks (100 per cent) are achievable for the average team. Alternatively, we can create an exam situation that mimics operational conditions and expect an error rate of around 20 per cent. I am in favour of the latter for one simple reason: we want to challenge and prepare our teams for real life. When we are searching for drugs, explosives, diseases, endangered species, rapist DNA, criminal tracks, or weapons, we can choose between finding nothing or an 80 per cent chance of success. Of course, we take the 80 per cent because we humans cannot equal the achievements of our highly motivated canine teams. For this reason, we should ensure they are well-equipped and motivated to give us their best – which also implies that we should trust them.

The more realistic the test, the better the dog will be prepared when he is on operational duty.

DEALING WITH NEGATIVE ALERTS

The criteria for false alerts are handled differently in both training and examination procedures. If a dog searches 20 cars and fails on two of them, does that count as a total failure and result in the dog being declared unserviceable? Or what if a dog incorrectly indicates on 5 per cent of the luggage at an airport? And how do we classify it if the dog misses two out of six hidden items in a search area? These questions must be answered in the protocol before training gets underway. Both trainers and dog-handler teams need to know what their goal is.

WHAT ARE FALSE ALERTS?

False alerts can be divided into two categories:

False positive: The dog alerts but there is no target scent.

False negative: There is target scent but the dog fails to alert.

Chapter 22:
PROBLEM BEHAVIOUR

What exactly is problem behaviour? This is not an easy question to answer because it is relative to experience, circumstances, and expectations. I have purchased a number of dogs for our unit from families who could no longer cope with their pets. Often they were desperate to find a solution after help they had sought from trainers or from vets had failed. The 'problem behaviour' generally involved excessive energy, attention-seeking, chasing, biting and resource guarding. In fact, it is precisely these types of dogs who have potential as scent detection dogs, and some of my recruits become highly successful drug and explosives detectors. We have also found many a good dog with 'problem behaviour' in KPNV dog sports clubs, where civilians prepare dogs for a role in the police. The exam regulations stipulate that a trainee protection dog must bark at a specific moment in the test and if they fail, they are excluded. But we love these 'silent' dogs because we can train them to work quietly as silent attack dogs on a swat team. What is a problem for one person can be winning the lottery for another!

I have also come across some very good hunting dogs who, as family dogs, were a disaster because of their desire to hunt everything that moved. They were not bred to be family dogs, but as hunting dogs it was easy to place them in suitable homes. They were given a job – to hunt – and the problem was solved. From these examples we can see that drive and energy pose a big problem for inexperienced guardians in a domestic setting, but in a different context, these characteristics are desirable and, indeed, sought after.

Before we discuss a 'problem behaviour', we need to analyse it:

- What exactly is going on?
- What behaviour is being offered?
- What is the intent of this behaviour?
- Can we use it, stop it, change it or ignore it?
- What will happen if we cannot prevent it?

We need to look at different aspects of the problem:

- How does the guardian feel about it?
- How do circumstances affect the behaviour?
- How does the behaviour affect the environment the dog lives in? A barking dog is very effective when trying to scare off a burglar, but a dog who barks all day in a rented apartment is a menace.

Let's consider the following examples of 'problem behaviour':

A woman asked for advice because her new companion dog kept biting her horse's legs when they were out on rides. This was something her previous dog had never done. The difference was that her new dog was an Australian Cattle Dog – bred specifically for herding cattle and keeping them in order by nipping their legs. This behaviour was in the dog's DNA. The 'problem' was a mismatch; the woman had chosen a breed that was unsuitable for her lifestyle.

Another owner asked for help because his dog bit him every time he tried to administer much-needed eye drops. The poor dog had blocked tear ducts and needed treatment three times a day. As soon as he saw his owner approaching with the medication, he started to growl and would eventually bite. He needed help, but it was too dangerous to treat him. His guardian feared the only option was euthanasia.

When a behaviour becomes so disruptive and dangerous that the only 'solution' is euthanasia – as with the dog who needed eye drops – it can be truly defined as 'problem behaviour'. Otherwise, it is simply 'unwanted behaviour' that can be influenced through training and/or management.

DOG DETECTIVE

It is important to bear in mind that every behaviour that is exhibited – even problem behaviour – has a function. Sometimes it is hard to discern – like the dog who keeps spinning and biting his tail – but, in fact, he is attempting to relieve stress. Of course, this only makes it worse – his tail may even have to be amputated at some point – but there is a reason for his behaviour. Sometimes a behaviour is easy to see, as in the case of the spinning dog, but sometimes it is so subtle as to be almost indiscernible. The heartbeat accelerates, muscles contract, breathing becomes more rapid, pupils dilate, as well as many other internal changes. Behaviour is on-going, even if we cannot see it.

Instead of labelling a behaviour as problematic, we need to identify the triggers that set the behaviour in motion – and what reinforces it – as this will allow us to make changes. Ultimately, our mission is to avoid euthanasia and provide a better quality of life for the dog, because the problem behaviour not only causes difficulties for the guardian or other stakeholders, but also for the dog himself.

Where does problem behaviour come from? There are many different causes but, before analysing it, we must first rule out a medical issue, as this is often the cause of disruptive and unpredictable behaviour. If health concerns are not implicated, we should look at the behaviour from the dog's perspective:

- Does the dog's living environment offer sufficient variety to avoid boredom?
- Does he have enough physical activity?
- Does he have sufficient mental stimulation?
- Is he constantly over-excited/over-aroused?
- Does he appear to be stressed?
- Could it be that a previous traumatic experience is affecting current behaviour?

DIFFICULT DOGS, DIFFICULT PEOPLE

There are no difficult dogs, and there are no difficult people. This is just label we put on a situation that is hard to handle. From a human perspective, this generally involves dealing with someone who is not doing what we want. Maybe it is not worthwhile for that person to follow our directions, philosophy or ideas. We, therefore, need to turn the tables and ask ourselves what that person might gain from co-operating. If we can offer something they would like, or which benefits them, the chances are they will do as we ask. We need to create a win-win situation for everyone involved – and that means seeing the world from the perspective of the people, and the animals, we want to train.

DEALING WITH PROBLEM BEHAVIOUR

Behaviour that causes problems comes in many different forms, but it is helpful to use broad categories to identify the type of behaviour we are dealing with, along with possible remedies.

AGGRESSIVE DOGS

There are many different types of aggression, and it can be directed towards other dogs, towards people or towards other animals. Broadly speaking it can be categorised as follows:

Fear Aggression: This type of aggression is usually in response to fear or a perceived threat. A dog may display fear aggression, such as growling, snapping, biting, or lunging, when he feels scared or cornered.

Possessive Aggression: This is when a dog becomes aggressive when defending his resources, which could be food, toys, his bed, or even his guardians. This type of aggression can be seen when a dog growls or snaps if someone tries to take away something he is guarding.

Territorial Aggression: This is triggered when a dog feels his territory is being threatened. He may bark or growl at a stranger entering the house, or when another dog enters his perceived 'territory'.

Protective Aggression: This is when a dog becomes aggressive when he is defending his family or his territory. He may growl or snap at a stranger or another animal that he perceives as a threat.

Social Aggression: This is directed towards other dogs or animals. In this situation, a dog may growl or snap at another dog when they are playing or when two dogs are fighting over a toy.

It is important to bear in mind that triggers are not always clearly identifiable and there can be underlying reasons such as an undiagnosed medical condition, the result of pain/discomfort, or the consequence of a traumatic experience in the past. Obviously, aggression can lead to dangerous, even life-threatening situations, and should always be taken seriously. Generally, it is not easy to resolve, and should not be attempted by the inexperienced. In cases of aggressive behaviour, it is essential to enlist the help of an experienced behavioural therapist.

SELF-SEEKERS

Many of the so-called 'problem behaviours' are shown because the dog wants to gain something. That is why we always try to find the answer to the question: *'what's in it for me?'* (see *page 49*) and look at situations from the dog's perspective. Consider the example of a dog pulling on the leash The reason for this could be:

- The dog wants to get to a certain tree because he wants to 'read' other dogs' urine markings.
- He may want to chase squirrels.
- He may pull towards other dogs because he wants to protect his guardian or claim a territory.

In such cases we are not talking about reactive behaviour but rather it should be seen as proactive behaviour (see *page 243*). The dog wants to gain something – reaching the desired tree, chasing squirrels or confronting other dogs – and pulling on the leash is his best means of changing the situation and getting what he wants. Observation plays an important role here. We need to find the triggers for the behaviour, looking for a specific context, place, circumstance, or a combination of these factors that prompt his response. In this way, we can collect data and write a training plan that is designed to cope with the problem behaviour. Remember, for the dog it is just behaviour; it is us humans that identify it as a problem.

THE IMPACT OF STRESS

We know from a number of wide-ranging studies that stress can be a major problem. It slows learning, affects relationships, and has a negative impact on mental and physical resources. The danger of slipping into depression is all around us, and the same can happen to our dogs. When staff in the emergency services are undergoing training, dealing with stress is a top priority. Mental resilience and physical strength have a positive effect on both health and quality of life, allowing us to deal better with the challenges we face at work and in everyday life. Consider the effect that physical activity has on our bodies. For those who love the gym, imagine how it feels to work out after a break? Think about the release of those hormones – endorphins, serotonin and dopamine – that make us feel good. Physical activity is a proven way to build resilience and to cope with stress – and the same goes for our dogs.

Stress cannot be avoided; it is an inevitable, everyday occurrence. This applies to all dogs: those living as companions, who cope with the hurly burly of their domestic surroundings, those who face the pressures of training and competition, and most particularly the 'professionals' who work in the police, military, customs etc. Consider the following stress factors:

Police dogs are often exposed to stressful situations. Most importantly, they need to be prepared so they are able to cope.

Living situation: This could be in a home or in a kennel. The dog may live alone or in a group.
Travel: Some dogs are driven in the back seat of the car, where they can see, hear and smell their handler/guardian, while others sit alone in a closed crate. In the case of service dogs, this could be for extended periods. Specialist dogs may have to travel by speedboat or helicopter when they are on duty.
Environmental conditions: Dogs may have to work/compete in varying temperatures, hot and cold. Service dogs may work have to work against a backcloth of gunfire and explosions.
Distractions: Focus on the task is required despite the distraction of spectators/bystanders, crowds at busy airports, train stations, etc, as well as the presence of other animals.

We demand a lot from our dogs, so we need to prepare them to deal with stress. For me, this means spending a lot of time with my dogs in addition to normal working hours. I never exclude them; I let them participate in my life. Walks provide important physical activity, but they also provide valuable mental stimulation. I turn them into mini discovery tours, where they are allowed to examine other dogs' olfactory information and where they can follow scent trails. I also play with them so we are engaging in a fun, shared activity. What we do in the home is equally important. Setting rules, such as feeding times and rest times, provides structure. Attention to detail ensures that they are well-fed and comfortable, and inclusion means they are mentally engaged and part of the family's activities. All these factors

IGNORING BAD BEHAVIOUR

When we witness 'bad behaviour' we have a tendency to ignore it. We bury our heads in the sand, look the other way, refuse to listen or to react. If we ignore bad behaviour, there is a possibility that it may disappear; we can manage the situation and not identify it as a problem. For example, if a dog raids the bins, we can simply place it out of reach. However, this strategy is not always feasible. What if the dog:

- Poops on the carpet?
- Jumps up at a small child who is holding an ice cream cone?
- Keeps barking while you are on the phone?
- Bites you because he wants the steak from your plate?

None of this behaviour can easily be ignored, and this applies to many different situations. In fact, ignoring behaviour should never be viewed as a viable response because we are responsible for our dogs' actions. Parents who have raised children quickly learn that wrongdoing cannot be ignored. We need to teach our kids that some things, such as stealing, lying or fighting, are not ok. As parents, we are both morally and legally responsible for our children; it is a duty we cannot escape.

The approach we take is less clearcut when it comes to our dogs' behaviour, and discussions can become as heated as when we are dealing with negative reinforcement and punishment (see *Chapter 17*). When it comes to bad behaviour, we can ignore a lot if the dog is placed in an environment where his wrongdoing has little impact, or may even be considered desirable. For example, a dog with resource defence issues and who is aggressive towards strangers, can live a useful life as a guard dog in a large, fenced compound. A Labrador Retriever who cannot resist hunting ducks or other wildlife can be trained to become a first-class drug detection dog, based at an airport. Problems arise if the dog risks harming other people, other animals, and even puts himself at risk. Therefore, the dog who tries to kill other dogs he meets on the street, bites small children, or attacks people, is displaying behaviour that should never be ignored.

improve quality of life and help to reduce stress.

DOGS WHO LACK MOTIVATION

There are dogs who lack the motivation to become professional working dogs or to compete in sports. The majority of dogs are relatively easy to motivate if it involves a game with a ball or tasty treats, and some respond to their guardians interacting with them or enjoying special attention. But with performance dogs and, most particularly, with professional working dogs, we need that extra motivation. Many dogs will be happy to use their nose and find target scent in a line-up, but most will lose interest after a trial or two. Many dogs will react with interest to a grasshopper jumping up in tall grass but spending hours searching for that one endangered grasshopper is another story. Almost all dogs like to play on a beach but looking for sea turtle eggs buried in the sand is not that exciting.

What can we do if a dog is not sufficiently motivated to carry out his allotted task? Ideally, a good testing system should be used to test this in the early stages of training. We use these for scent detection, tracking and attack dogs, as well as for other specialists. But even if we have selected highly motivated dogs, we may reach a stage in training, or in the workplace, where we are confronted with a lack of motivation. Our first response should be to check the dog's health, assessing his body temperature, his condition, and his general behaviour. The next step is to look at the training data records. Mostly, we find that motivation starts to wane a week or two before we realise it. If we are working with an intrinsically motivated dog who is declining in motivation, we need to look at attitude, environment, success rate in training, as well as his mental and physical condition. On the other hand, if we work with dogs who are more externally motivated, we will also need to look at the reinforcement rate, the value of the reinforcers, or the effectiveness of the negative reinforcers.

In addition, we need to check the environment to see if there are any distractions that we humans have missed. A bitch in heat? A high-pitched noise? What could be the problem? Often, a drop in motivation is because we have made a mistake: the reinforcement rate was too low, the reinforcer was not sufficiently interesting, the exercises were too difficult, or maybe there are problems in the relationship between dog and handler.

In these situations, we need to scrutinise every detail that could be relevant, which involves checking health and training data, and working with positive reinforcement. We need to build positive expectations, trust, and a desire to participate in our games, which may require criteria to be lowered.

DOGS WHO SELF-REWARD

The key element here is that the dog is in control! He has found a way of behaving that reaps rewards – for example, raiding the kitchen bin for scraps of food. As we know, behaviour is a result of its consequences, which means the consequence keeps the behaviour going. This is something we cannot ignore because the dog will continue to reward himself. To stop the behaviour, we must eliminate the reward. For example, we get a secure lid for the kitchen bin, or we put it in a place where the dog cannot access it. Once the potential to self-reward has gone, the behaviour will go away.

It is important to distinguish between a dog who self-rewards, and behaviour that is, *in itself*, rewarding for the dog, i.e., it does not require external reinforcement, such as the food from the kitchen bin. The behaviour is intrinsically motivated. We have understood from Skinner that behaviour which is reinforced is repeated and behaviour which is not, dies out. However, intrinsic motivation is different; it is something from within that makes the behaviour stronger. For example, a Border Collie does not need an additional reinforcer to make him herd sheep – the activity alone is reinforcing for him. This applies to many everyday behaviours: running, chasing, marking, barking, digging, retrieving and sniffing are all examples of intrinsically motivating behaviours for the majority of dogs.

This brings me to the question of whether it is necessary to reinforce a scent detection dog following a search where nothing is found. Handlers often use a '*well done*' or an '*all clear*' signal after a 'empty' room search and then use external reinforcers to reward their dogs. However, this can cause problems if the dog learns that mimic searching, or not alerting on a hard-to-locate target scent, will also lead to a reinforcer. In my opinion, most scent detection dogs are selected because they are intrinsically motivated to use their nose. So, if my scent detection dogs failed to find target scent in a room, they would be only too happy to go to the next room and try again. The opportunity to restart the search is sufficient motivation, and there is no need for an external reinforcer.

UNWANTED BEHAVIOUR

Initially, a dog has no concept of either wanted or unwanted behaviour, and he cannot tell the difference between them. It's all about how worthwhile the behaviour is for him. The learning curve is that others around him – other dogs or humans – perceive a behaviour as unwanted. It is the consequences of this response that will affect his on-going behaviour.

There will always be times when a dog decides on a course of action which his guardian cannot influence at the onset. For example, the dog may decide that nobody is allowed to sit on the sofa that he claims for himself, or he may continue barking at other dogs and people, even when he is told to stop. If boundaries are not clear and the dog takes control, we have a serious problem. It is obvious that there are issues involving hierarchy and dominance, although these terms are much frowned upon. So let me explain in a little more detail what I mean.

If a dog barks aggressively when a visitor comes to the house, he is not showing dominance. Dominant dogs are very calm, confident and survey the situation. Dominant dogs decide who eats first, they claim the best spots in the house, they dictate the behaviour of other dogs in the house, and they may – but not always – try to control people as well. Dominant dogs may also be protective in their behaviour.

Hierarchy is something we have all experienced in our lives. It starts at home with our parents deciding when we should go to bed, what we should eat, how we should spend our holidays, etc. It continues at school and college where teachers set rules and hand out assignments, through to job situations where the boss /team leader is in charge of what we do. In the military, police, and other large organisations, we are well acquainted with a chain of command.

A hierarchy can be seen as a structure that provides both stability and accountability. It provides a framework, controlling the environment, assigning tasks, and resolving conflicts. Hierarchies ensure 'survival' and balance and our dogs rely on us to provide this structure to their lives.

REACTIVE DOGS

A reactive dog is often described as one who barks at other dogs, people or cars. But dogs that pull on the leash, or pounce on other dogs, are also often described as reactive. But is 'reactive' the right description for these problems? Reactivity occurs when individuals respond to each other by changing their behaviour. In dogs, reactivity is generally perceived to be a negative response but it can also be positive. A great example of positive reactivity is the so-called Pygmalion effect where pupils change their behaviour to meet the teacher's expectations.

Positive reactivity also has a part to play in the training process. For example, a handler is told that he/she has a really good scent detection dog. As a result, their approach and their interaction with the dog becomes more positive. This is picked up by the trainer who creates special training opportunities and spends more time working with this combination compared with the other dog-handler teams in the group. They are chosen to do demos and recruited for special operations where they gain experience and become still more successful, such is the power of positive reactivity!

There is a definition of reactivity that appeals to me:

"Reactivity is a term used to describe a substance's tendency to undergo chemical change in response to the presence of other substances. It is a measure of how quickly a substance will react with other substances or how easily it will undergo a chemical reaction."

This leads me to the question of whether dogs can be considered 'reactive' when they are releasing energy (pulling on a leash, barking, etc.). As trainers, our priority is to find out *why* a dog is being reactive. Why does he pull on the leash every time he approaches certain bushes on his walk? Why is he chasing squirrels? Why is he aggressive towards other dogs in the park?

There is no standard response because every dog has his own personality, his own drives and his own life experiences. We should also bear in mind that it is not only the dog but often the dog-human combination that triggers a reaction. This can be put to the test when, for example, someone else walks the dog and there is no sign of reactive behaviour.

PROACTIVE DOGS

Being proactive means creating or controlling a situation by setting events in motion, rather than reacting to something that is happening or has already happened. So, to identify and prevent potential problems, we have to be proactive. Proactivity also makes a positive difference in the workplace, as it encourages efficiency – more is achieved while expending less energy. Proactivity is better than reactivity and can even save lives. Most safety precautions are proactive in nature to prevent accidents; roping in when working at height, wearing a helmet on a construction site, or wearing a seat belt in a car. Even the reactive fire department spends a lot of time and money on fire prevention.

So, it's no coincidence that when I am recruiting new personnel for our unit, I look for proactive humans, and I also look for proactive dogs – those who love to use their noses, jump, pull, chase, bite, attack, etc. We use their drive to support us in our work or in competition. But these proactive dogs can also cause problems if they are not sufficiently challenged, if they spend most of the day indoors and only get a few minutes' exercise. They will expend their surplus energy on random occupations, such as picking up anything they can find in the house, herding children, stalking the family cat, or over-engaging with visitors. Mental and physical exercise is essential for proactive dogs.

There is another factor to consider. We all prefer *proactive* people; they are the problem-solvers, and they get things done. In contrast, *reactive* people are more likely to complain than attempt to resolve an issue. This terminology has a powerful effect. Unconsciously we condition ourselves and create associations. So, when we describe a dog as being proactive rather than reactive, we see situations in a different light. We perceive the dog's actions with a more positive energy which, in turn, make us better trainers.

We need to enjoy our proactive dogs. We need to work at understanding what drives them and adjust our training accordingly. We need to make sure there is enough activity in their lives and to appreciate their 'can do' personality!

A proactive dog is waiting for training at the line-up to get underway.

THE NATURE OF LEADERSHIP

Dogs either follow leaders or are leaders themselves. This is a mantra that I wish more trainers/guardians would follow because, all too often, dogs are put in positions they don't want to be in, and it throws them off balance. This is when energies shift and behavioural problems surface. It could be the toy dog who is carried around in a handbag, the large dog that pulls ahead, barking at every dog he meets, or the dog at the front door, barking furiously at visitors. Many behavioural problems are related to leadership problems. Unfortunately, as soon as we address this, we get drawn into discussions about hierarchy and dominance (see *page 68*).

There is a hierarchical structure in both human and canine society in which certain individuals have authority over others. However, the nature of the hierarchy and the roles of each individual within it may vary between species. For example, hierarchy among humans is largely based on social status, whereas hierarchy among dogs is based more on dominance. In addition, the degree to which hierarchy is present within each species may also differ. In humans, for instance, hierarchy may be more rigidly enforced, whereas in dogs the degree of hierarchy may be more fluid. The implications of hierarchy for each species, such as the potential benefits and drawbacks of such structures, merits further discussion.

At this point, I would like to emphasise that I am completely against the traditional methods of dealing with so-called dominance - for example, pinning the dog down to exert authority. I have never seen a boss pin down his employees until they no longer resist. Leadership is much broader than exercising dominance. It starts with exuding calmness - taking on the role of a calm, confident, and proactive leader. We are there to give directions, to build confidence, to solve problems, and to stand up for our dogs - whatever happens. We are not exerting our authority by being harsh disciplinarians; our aim is to be perceived as being both trustworthy and dependable.

One of the ways this can be achieved is by setting the right example. If we are asking a young, inexperienced dog to negotiate a metal staircase, for example, we should lead the way, if we want the dog to jump from a wall, we should jump first. A calm and confident leader not only shows the way but goes first. I am not advocating the *'do as I do principle'* (see *page 73*). I want dog and guardian/handler to work as a team, to support each other, help each other and trust each other. Inevitably, there will be times when a dog becomes over-aroused. A calm and proactive leader can absorb that excitement and control the energy. It's all a matter of mental attitude. If we can control our state of mind, we can influence the dog's emotional barometer. It is this calm energy that makes a difference when interacting with both dogs and humans.

For example, a leader should remain relaxed and composed in difficult situations, demonstrate empathy and understanding towards others, and focus on solving problems rather than getting angry or frustrated. When working with others, a leader should maintain a respectful tone of voice, be open to different perspectives, and take the time to listen to the concerns and ideas of their team. Additionally, leaders can use their calm energy and attitude to encourage collaboration and foster trust among their team members. This calm energy will also calm our dogs. If we prepare ourselves for conflict, our dogs will tune into this and react accordingly.

We therefore need to ensure the dog stays calm before he gets his food, before he is let out into the garden, when someone is at the door, or when visitors are in the house. He should learn to step aside when coming through doors, rather than pulling ahead, to stay in a 'sit' or a 'down' while his food is being prepared, and to wait for permission before he is allowed to eat. In this instance, we can add duration, and distractions to strengthen the behaviour. In our living room, our dogs each have their own bed. If we have visitors, we send the dogs to their beds, a place where they feel comfortable, and we can entertain our guests in peace. If the dogs remain in place, lying on their beds quietly, I reward them with chews.

This is very different to excluding dogs from their human family, either sending them to another room, to their crate, or to a kennel for extended periods. This form of exclusion can lead to mental problems and is often a trigger for unwanted behaviour. We are not always aware of this, and yet it is so easy to avoid. We need to make a list of the day's activities and look for opportunities to involve the dog. My dogs love to go hiking with me, but they also love to lie next to me when I read a book. They watch me when I am pottering about in the kitchen, and we enjoy each other's company. Simply spending time together is important for everyone.

FINDING THE BALNCE

All dogs need exercise and the opportunities to be active. This is easy to do by extending daily walks and exploring new places. Many dogs also like to run next to a bike, swim, hike or play with a ball. This is not only good for the general health of the dog, but also for his mental equilibrium. In addition, it also strengthens the bond between dog and guardian. If it suits the dog/guardian combination, negotiating obstacle courses can be great fun, exercising mind and body, as well as honing basic obedience/handling skills. The course can be challenging, making use of a range of equipment, or it can be simply a matter of engaging with the environment – finding low walls to jump on and off, tree trunks for balancing and narrow gaps to crawl through. Getting involved in dog sports is a great way to keep dogs busy and active, and there are plenty to choose from – agility, flyball, obedience, scentwork, search and rescue

Dog sports, such as agility, are a good way to provide physical exercise and strengthen the bond between dog and guardian.

WALK THE WALK

One of the fundamental skills I work on with all guardians/handlers is the walk. Once we master the technique of walking a dog, we set the tone of our relationship. Again, this is simple, but not easy. The aim is to walk the dog on a short but slack leash, making sure he walks alongside his guardian/handler or slightly behind. The guardian/handler should be proud and confident: upright in posture, with shoulders back and arms relaxed so there is no tension in the leash. We should also give the dog the opportunity to interact with the environment, engaging his senses, and taking in what he hears, sees, smells, and touches. He needs to be able to look around, listen to sounds, smell other dogs' urine marks, taste things, and roll in the grass. Structure in a walk and guidance from the handler are important, but so are the breaks where the dogs can do as they want and interact with the environment.

and many more. Working together strengthens the relationship between dog and guardian and takes it to another dimension. Everything revolves around energy. If the guardian is asking too little, and the dog has excess energy, a conflict will arise. We need to find a balance so that the energy flow is positive.

When dogs have sufficient exercise, both physically and mentally, many behavioural problems go away without the need for further intervention. The dog stops attacking other dogs, and people. He no longer barks or becomes destructive when left home alone. Instead, he sleeps because he is happy, balanced and relaxed. He is settled because he knows he has abundant opportunities to expend his energy and find mental stimulation.

Dogs are dynamic creatures and love to run, jump, crawl and roll. But our everyday lives are full of appointments, work, children, television, internet, shopping, cooking and other tasks or distractions, and it is often the dog who suffers. We, therefore, need to examine our daily schedules and work out how much time we have left to spend with our dogs. And what about training? Do we still have the time and energy left to work with our dogs, whether it be in basic obedience, a few fun tricks or getting involved in one of the sporting disciplines? The average household allocates around 30 minutes a day exclusively to the dog. This is not enough, and it will, inevitably, lead to unwanted behaviours, as already outlined. When attempts are made to resolve the situation using punitive methods, things go from bad to worse. As already highlighted, in order to solve a problem, we first need to know what has caused it. Too little exercise is one of the main causes of unwanted behaviour, and lots of exercise is one of the best solutions. Dogs are born to run and explore; they need mental and physical challenges. Going out for 10 minutes three times a day is simply not enough!

Varied exercise and mental stimulation are key ingredients for a balanced mind.

Chapter 23:

INSIGHTS INTO HUMAN BEHAVIOUR

In my seminars when I discuss working with trainers, I always say that training animals is the easier part. The bigger challenge is working with people because issues such as integrity, principles, shame, guilt, ego and conflict can all play a part. However, my work with animals has given me insights into human behaviour, and I have discovered that there are parallels between training animals and managing people. In this chapter, I would like to share some of these insights.

THE IMPORTANCE OF PRINCIPLES

Principles are guidelines that help us navigate our lives. We live in a fast-paced world with lots of distractions and sometimes it's hard to keep up to speed. There are moments that require immediate action, and moments when we have no idea what to do. In these pressurised situations, integrity, and relying on principles, can help us to make the right decisions, Others need to know who we are, what is important, and what we stand for. In this way, we attract listeners who also identify with our principles.

As I write this, I am sitting on a KLM plane high above the Atlantic Ocean. I always book on this airline because of the way they do business. One of their principles, for example, is to offer customers high-quality organic food and to serve it with plates and cutlery made from compostable materials. The staff is trained to follow the airline's principles so that customers know exactly what to expect, which is quite an achievement considering KLM employ over 50,000 people in constantly changing cabin crews. Nevertheless, the staff work together as smoothly as a tight-knit team that spends every working day together. The principles are clear for everyone: the crew, the pilots, and also the customers who must observe a standard of behaviour, i.e., not abusing staff or being drunk on a flight, in order to fly with the airline.

Principles guide us, but they are not absolutes. There will always be moments in life that surprise us and present us with new dilemmas. These are precisely the moments when we need to add to our list of principles or sharpen their description. As individuals, we should take a moment to reflect on our own principles:

- What is important to us?
- What could help us stay on our chosen path?
- How do we deal with distractions in life?

We need to work out the salient points and see if our words tie in with our actions.

APPROVAL RATINGS

Most of us seek approval and we thrive when our efforts are met with appreciation. We want to be seen, heard and recognised – it's in our DNA. The way we express ourselves has a lot to do with seeking approval. It can be the way in which we speak or write, or the way in which we move. It can express itself in sports, hobbies, work or interpersonal

Validation, and receiving credit for work well done, are important for both human and dog.

relationships – we seek validation in many different ways. Appreciation fuels our egos, and this applies to everyone: those with big egos, those with small egos, those who flaunt their egos, and those who are more discreet. When we win approval and our work is appreciated, we feel good. We feel more confident, and that motivates us to try harder and to be more inventive. In contrast, if our work is not acknowledged, we lose drive. We stop trying to improve our skills and give up in the face of difficulty or disappointment.

But, of course, the ego also has its dark side. When a person's ego becomes inflated, and focuses solely on getting approval, there is a danger of resorting to underhand measures such as stealing the work, ideas, knowledge, and intellectual property of others. Obviously, this weakens trust, but the victims also miss out on getting credit for their work; their contribution is under-valued, and their self-esteem suffers. In the business world, this lack of scruples is often motivated by a desire to generate more profit; in the music industry, it may be driven by a desire to gain popularity. In the world of animal training, successful outcomes, or otherwise, can make or break a trainer. Economic survival may depend on doing a better job than others, achieving results faster, or being able to train 'difficult' animals. Of course, ego is also at work here but there is nothing wrong with that if we are also

PRACTISE WHAT YOU PREACH

Our words are not always in line with what we do or don't do. Almost every day we hear of politicians or CEOs of big companies who have crossed the line and failed to act with integrity. Clearly, the gold standard is to practise what we preach. Sadly, I have seen too many videos, taken undercover, which expose police dog handlers using punitive measures, and I have seen highly respected dog trainers telling their audience that their work is based on 100 per cent positivity, when in reality they use negative reinforcement and punishment as well.

To reiterate, our overall focus is to use positive reinforcement wherever possible. However, as previously outlined, negative reinforcers and punishment are a part of everyday life (see *page 50*), so it is hypocritical to claim to work 100 per cent positively and then be unable to implement it. Theory has its place, but we need to live in the real world and find solutions that work for the guardian, the handler, the trainer, the, operation/competition and, perhaps most important of all, for the dog.

discovering new and better ways to train and share our knowledge.

There is enough work for everyone in the world of dog training. Sure, if your name is out there, your business may grow faster and you may make more money than others. But truth prevails. Acknowledging the work of others – thanking mentors and teachers – sends an important signal to clients, as integrity is an important factor in deciding whether to hire someone, or whether to follow a particular training method. Trainers who are prepared to share information demonstrate that they value each other, and are happy to acknowledge the help they have received. This allows the training community to grow together, to discover new things, to be innovative and to improve the lives of both dogs and people.

THE ROLE OF MANAGEMENT

Managers can learn a lot from dog training because there are many parallels. Managers must deal with wanted and unwanted behaviour, impose boundaries, set clear criteria and offer rewards, but punishment also has a part to play. Humans are nothing more than highly sophisticated mammals. We are driven by pain or pleasure, i.e., we want to avoid pain/suffering and we seek pleasure/wellbeing. These driving forces are very clear and simple; it becomes more complex when we focus on the individual. It's not just our differing personalities; context, experience, fatigue, timing, emotions, and many other circumstances also play a role in how individuals perceive reward or punishment. This means that, like animal trainers, managers/bosses need to know the individuals they are working with. They need to know what motivates them, they need to communicate clearly with them, and set achievable goals.

This parallel has been tested by getting managers to work with animals. I have trained company managers in operant conditioning using chickens as live training models (see *page 74*). In addition, there are equine coaching programmes, management workshops with dogs and even one where monkeys are used. Typically, a professional coach works with a manager, and guides them through a series of activities involving the animal. The manager learns how to read the animal's body language and how to respond to the animal's behaviour with the aim of building a trusting relationship. The coach provides feedback and guidance to the manager on how to best interact with the animal and adjust their approach when needed. This type of coaching programme helps managers learn how to better control their own emotions, build relationships, and improve communication with team members.

When I visited Appenheul, a zoo of free ranging primates in the Netherlands, I was given special permission to join a gorilla group in their living quarters. Before I was allowed into their space, the head of the gorilla group first checked me out. I had to step close to the fence, allowing the gorilla to engage in intense eye contact and to smell me. Only then did he give me permission to visit his living quarters. I learnt a great deal from this encounter.

Unfortunately, many companies still operate on the principles of the 19th century, i.e., employees are the workers and managers are the thinkers. Problems arise when managers decide that workers are not sufficiently productive and use

LEARNING FROM CHICKENS

My company offers business leaders an opportunity to work with the principles of reinforcement that have been proven by behavioural experts, psychologists and scientists such as Thorndike, Pavlov, Skinner, Premack and many others. In the seminars, we study the principles they set out, such as criteria, rate of reinforcement, communication, limits, training schedules, data collection, and so on. And to make sure managers can apply this theory, we let them work with animals. A group of dogs for management training might be difficult to manage, so we use chickens instead. They are easier to deal with in the seminars and, most particularly, they are very fast and pay close attention to body language, so the trainers must be crystal clear about what they want to reinforce, and exactly when they should mark the behaviour. Chickens do not work under stress, threats, or punishment. Equally, complex behaviours can only be achieved if trainers follow a plan, break the behaviour down into small steps, establish clear criteria, collect and evaluate data and work with positive reinforcement.

This type of management training is intensive; the days are filled with theory and practice, and in the evening the participants are given the task of evaluating the data they have collected and drawing up the training plan for the next day. There will always be competition between the participants and, inevitably, there will be disappointments – but also 'aha' moments that bring great joy. People are taken out of their comfort zone, the relationship between the participants improves and at the end there is a test where participants showcase their chicken training. This is often exciting and emotional – something they will never forget!

Chicken camps provide an opportunity not only to practise training skills, but also to test management skills.

coercion and punishment to motivate them. In addition, managers struggle to communicate and fail to set desirable and/ or achievable goals. No data is collected and analysed to verify results and there are no protocols or guidelines and, perhaps most important of all, there is no vision.

LEARNING TO COMMUNICATE

All of this leads to poor communication, criteria that is unclear and often a low reinforcement rate. In many companies, managers devote around 80 per cent of their time to problem-solving. How much better would it be if they devoted 80 per cent of their time to adopting a positive approach when dealing with their workforce, rewarding employees on the basis of performance and achieving goals? Managers need to act like coaches and create an environment where everyone can thrive, learn new skills, and work towards a shared goal. Just as with animal training, we have to set boundaries, use clear criteria, set employees up for success and reinforce wanted behaviour!

I have come across some fascinating data which showed that almost 85 per cent of mistakes in the workplace are caused by procedural guidelines that were unclear and/or potentially dangerous. Lack of skill accounted for 14 per cent of errors, and one per cent was intentional. Good communication can make a huge difference. This is backed up by a study conducted by American psychologist, John Gottman, who found that the companies with more positive communication were by far the most successful. Gottman speaks of the magic 5:1 ratio in communication, which means that one negative interaction must be followed by five or more positive interactions, both in the workplace - and in a happy marriage! And I strongly suspect that this applies to animal training as well.

Clear is friendly, unclear is unfriendly - the art of communication really is that simple. We have learnt from animals that we must be clear. There is reinforcement or there is not - there are no grey areas. The same applies to imposing boundaries. We need to ensure predictability. If we are unclear, that is unfriendly. Are dogs allowed on the sofa or in the kitchen? There is only a 'yes' or a 'no' and it must be clear. This degree of clarity should also exist between a boss of a company and their employees. For example, if the private use of company cars is not permitted, this must be clearly communicated. Then everyone knows the rules - employees, team leaders and CEOs - and that is fair and friendly.

Reinforcing wanted behaviour among employees should be the top priority for supervisors and team leaders. We must be clear about what is wanted instead of focusing on failure. Self-esteem suffers if the onus is always on what has been done wrong - and motivation fades. Soon the desire to try disappears for fear of making mistakes which, in its extreme form, leads to a state of 'learned helplessness' (see *page 135*). This phenomenon sometimes comes to the fore in meetings when, all of a sudden, no one is participating. Everyone is silent and no one comments on the speaker's remarks or suggests ideas - a clear sign that this is not a safe work environment and management should be questioned.

Animal training can help managers to recognise and avoid these pitfalls because, as we know, it is unclear criteria, unclear boundaries and too little reinforcement that lead to problems. The use of sanctions or the threat of sanctions does not help. On the contrary, it will poison the company from within. An unsafe environment is not a good work environment, and once management struggles with trust issues, they are very difficult to overcome.

If 'punishment' is off limits, what course of action can a manager take? Working with animals is a real eye-opener in this situation. Take the example of the dog who fails to sit on cue:

- Did the dog understand the cue?
- Does he trust his trainer?
- Is he distracted by anything?
- Is he unsure about the skill?

In all these scenarios, punishment would be devastating. The solution is to offer help, be clearer on the criteria, and use a high value reinforcer or a higher rate of reinforcement. Then we not only get better behaviour, but also a happy dog who has a good relationship with his trainer.

The idea for these intensive chicken workshops came from Marian and Bob Bailey, who were the first to run them, initially for their own animal trainers. I saw enormous potential in this for everyone who works with animals, but also for those working at management level. Later we ran the chicken workshops not only for animal trainers and managers, but also to select new training candidates for our government contracted animal training projects. Today I share this knowledge in my own company to open up new opportunities for as many people as possible, even if they may not have anything to do with animals in their everyday lives as managers - but they benefit enormously from it for their professional work!

Chapter 24:

LEARNING FROM SCIENCE

The scientist observes, examines the evidence, and tries to find answers. In the realm of dog training, the scientific approach would be to ask the following questions:

- Why do dogs work for us?
- What exactly can a dog see?
- What frequencies can he hear?
- How does he process scent?

This line of research helps us to become better trainers.

Interestingly since 2005 there has been a noticeable increase in scientific publications that focus on dogs' sensory abilities and provide new insights regarding their cognitive abilities. This awesome research helps us to understand our dogs better, and to train them more effectively. This, in turn, leads to enhanced performance and successful outcomes on ever more challenging tasks.

REVISIONS START HERE

To name a few outstanding examples of scientific research:

- Dr Nathan Hall and Dr Paola Tiedeman, based at the Texas Tech University (TTU), have worked on several collaborative research projects merging analytical chemistry and canine behavioural assessments. Chemical work was geared towards the identification of odour volatiles emanating from a range of explosives, as well as evaluating environmental conditions on volatile emission and movement processes. Understanding how the odour profile changes under extreme environmental conditions has helped to promote better canine training regimens and the odour quality from regularly used explosive training aids.
- On a parallel level, canine behavioural assessments from these collaborative efforts have been vital in understanding how environmental conditions affect canine physiology, and how training regimens under distinctive climate parameters can mitigate drops in performance and better inform training in operational challenges. Results have shown that a decrease in detection rate can, for the most part, be attributed to the biological/physiological effects on the dog, rather than the odour availability. This was exemplified by performance levels dropping in high temperature conditions, especially in high temperatures combined with high humidity, where odour availability was at its greatest. This work highlights the potential that canine acclimatisation and conditioning could have on mitigating potential performance decrements. They have now expanded their research to cover a wider range of explosives, different training paradigms and continuous monitoring of odour availability and persistence.

In addition:

- Dr Brian Hare has examined how dogs learn and how they relate to their guardians.
- Gregory Berns has used MRI scanners

to see what is going on in a dog's brain when he gets cues that lead to reinforcements.

- In their book, *Canines: The Original Biosensors* (published 2022), Dr Lauryn E. DeGreeff and Craig A. Schultz commissioned a diverse line-up of writers from a variety of fields to show how scientific knowledge can have a major impact on understanding and utilising the dog's olfactory system.

Scientists have always seen the animal kingdom as a major resource and they have gained valuable information from studying anatomy, observing how animals move, how they make use of their natural skills, and how they employ their brains in the decision-making process. This knowledge can them be transferred to other spheres. For example:

- The extraordinary aerodynamic anatomy of birds has been the model for designing aeroplanes and helicopters.
- Spot', the robot dog, developed for exploring dangerous//inaccessible areas and buildings, was invented following detailed studies of canine anatomy and movement.
- The navigation abilities of bees have been used to develop autonomous robots and drones.

Joining forces: Police dog Ops and robot dog Spot, designed by Boston Dynamics.

The study of the structure and form of organisms, as a subfield of biology, is called morphology, from the Greek words 'morphe' meaning shape and 'logos' meaning word or knowledge. With the help of morphology and so-called bioengineering, which takes models from nature for new developments in technology, scientists have been able to help us to live and work more effectively, more efficiently and in safety.

This could not be achieved without the help of trainers who making testing possible in a variety of ways including:

- Working with birds to ensure they stay in the same position in a wind tunnel so that scientists can film how their wings cope with different wind speeds.
- Training dogs to do cognition tests so scientists can learn more about their memory and decision-making.
- Teaching animals to move from one place to another place so scientist can track the way they move.
- Testing sealions by covering their eyes as they hunt fish so scientists can assess how they use their whiskers.
- Training dolphins to scan through black boxes and do match-to-sample exercises so scientists can study how they use their sonar.

The benefits of trainers and scientists working together are tangible. The problem is that scientists and trainers often speak different languages. Scientists are in search of proof and facts; trainers like to follow their feelings. Scientists are deterministic, trainers are stochastic. Scientists want to collect data and follow set procedures; trainers just want to train. We can all benefit if we are curious, listen to each other, and are prepared to help each other.

If asked to define the nature of science, scientists will give a range of answers. I believe it is best described as: *the knowledge of the physical or material world, acquired systematically through observation and experimentation.*

However, this tells us that not everything can be explained by science. For example, in the world of dogs and dog training, a lot of things are happening that we have yet to fully understand. American astrophysicist, Neil de Grasse Tyson puts it so well:

"The good thing about science is that it's true whether or not you believe in it."

Looking at animal training from a scientific perspective, there is no such thing as 'knowledge' or 'truth' until a specified behaviour can be replicated by many other animals, trained by many different trainers. Scientists note every little detail of their experiments so their work can be peer reviewed or used in comparative studies. We, as animal trainers, are not in the same league, in this regard. A tiny minority of trainers record data systematically and seriously, and fewer still are willing to share their training protocols with others. And I have never known a trainer to ask other trainers to check their results or replicate their process to see if they get the same result. We could learn so much together! Scientists strive for objective results that lead to objective truths. As animal trainers, we could benefit from this approach, especially when it comes to writing clear plans and protocols and keeping bias out of our training and experimentation.

Scientific research is not confined to animal physiology; there is a growing interest in animal behaviour. As I highlighted in the previous chapter, studying animal behaviour can help us better understand basic principles of human behaviour. By understanding how animals think and act, we can gain valuable insight into our own behaviour and reactions. We can understand the cause and effect of our own behaviour, and this can lead to the development of new therapies and treatments for mental health.

The way in which animals learn has also been a fruitful area of research extending back over some 150 years. How animals maximise the use of their senses, how we can teach them to deal with stress, how best to reinforce, and much more. In my own field, the input of scientifically proven information has had a major impact on the ways dogs are trained, and how they are treated.

In the 1990s, if a scent detection dog made a false positive indication, he received physical punishment. If a tracking dog lost the track during training, he received physical punishment. In those days, an active alert - barking and biting at the source of the target scent - was mandatory because it was thought that dogs would not work if an element of obedience was introduced. And if a dog failed to give an active alert when target scent was present, he was punished for not doing his job.

Science helped the K9 community to overcome these paradigms. For example, scientists have explained why a tracking dog may lose a track, and why punishment will exacerbate the issue. We now understand that, in some circumstances, when there are specific combinations of temperature and humidity, it is impossible for dogs to detect the target scent. In addition, scientists have shown us that obedience will not influence the work ethic of a scent detection dog, and they have helped us teach dogs a wonderful passive alert with a positive reinforcement approach.

So, the best advice I can give to animal trainers is: make friends with scientists! Approach each other openly, work together, be curious, speak each other's language and learn from each other.

SCIENCE VERSUS REAL LIFE

Of course, there is a difference between an experimental set-up in a laboratory and real life, for example in a police/military operation. In laboratory conditions, we can test how quickly a dog reacts to given cues or where the perception threshold for certain odours lies. But if we use the same explosives detection dog in a war zone in Afghanistan, where it can be very cold at night and extremely hot during the day, the results would be different, especially with the distraction of other people and animals. The dog may even be operating under fire. But despite the difference between the laboratory and real life, there is still much we can learn from science.

In my 25 years in special operations, I was involved in a variety of projects and studies where scientists and trainers worked to solve operational puzzles. I wish I could give details of the work and credit those who were involved, but I am unable to do this for security reasons. However, what I can share is that these studies focused on helping us humans understand more about the capabilities of our dogs. We introduced dogs to changes in air pressure, temperature and humidity to monitor their reactions and see how they would cope with working in more

GARBAGE CAN SCIENCE

It was Bob Bailey who taught me the value of 'garbage can science' in bringing ideas to life. Together, we worked on developing new instruments and aids for training. However, building prototypes was expensive and time consuming so we often conducted tests using a simple model. We called it garbage can science because the materials we used were often literally from the trash – scraps of cardboard or wood held together with Duck tape, string, or staples. We let the dogs work with the model to see if the principle worked. When it was thoroughly tested, we asked our engineers to make the device in all its complexity because we knew it would be worth the investment of time and money!

A vertically mounted scent wheel, built courtesy of garbage can science. The wheel spun while the dog stood in front of it and sniffed out the various scent samples. Reinforcement was provided by ball machines attached to the wall and to the device.

extreme conditions. We worked in rooms lined with video screens to mimic real life scenarios, and we introduced noise and other distractions to test them further. All these studies were carried out under ethical jurisdiction; the dogs' welfare was prioritised, and all testing was based on positive reinforcement. The results were mind-blowing in terms of what we learnt from this, and how we could train our dogs to be fully operational in the most extreme situations. With the help of scientists, we are changing fast – we are not only changing our opinions, we are changing the way we train, and the way we work with our dogs. The way we organise our odour recognition tests (ORT) today is a good example of this (see *page 229*). Yet there is still so much to learn, which we can discover by sharing knowledge and experience.

TARGETING SCENT DETECTION

In my own field of scent detection, we have been working with scientists and made some important discoveries.

In recent times, we have been focusing on the most effective method of releasing target scent. Odour release is determined by the quantity of odour and by vapour pressure, which affects the volatility of a substance and its ability to be released into the atmosphere. The amount of odour determines the intensity of the odour released. For example:

- A substance with high vapour pressure but low odour concentration results in a less intense odour being released into the atmosphere.
- A substance with lower vapour pressure but higher odour concentration results in a more intense odour being released.

In addition, we need to understand that vapour pressure will, eventually, fill the open space – for instance a drawer, a box, or a cabinet. Therefore, if we are working with a substance that is very volatile, it makes no difference whether we hide 10 grams or 1 kilo of cocaine because the airspace becomes saturated with odour molecules. Adding more cocaine/target substance will not provide the dog with more odour.

We need to be aware of the impact of odour delivery devices (ODDs) and Odour Carriers have on training. Imagine working with a toxic targets, such as poisoned meat that is used to kill birds of prey or dangerous chemicals, or imprinting a dog on endangered animals, such as small lizards or frogs. In these scenarios, we need to ensure that the dog never comes in contact with the target. An odour delivery device (ODD), which could be a metal container, or some form of protective material, will prevent contact. They can also protect target scent against the influence of weather conditions, such as UV in sunlight, rain and temperature differences. But there is also a downside. A dog can learn to recognise the ODD by its shape, colour and material by using his nose and/or his eyes. Therefore, when using an ODD, we need to provide distraction odours in ODDs and use empty ODDs to make sure the dog is only searching for the target odour.

In addition to ODDs we can also make use of odour carriers, odour absorbents or soaks. These are different materials that are used to carry or absorb the target odour or scent. For example, if we place a drop of a liquid target odour on a piece of plastic it will retain its shape. However, if we use a piece of paper or wood, it will soak in, increasing the surface of odour release, thereby affecting the overall odour release. Like ODDs, odour carriers, absorbents or soaks can protect target odour against the influence of weather conditions, such as UV in sunlight, rain and temperature differences. However, they can also influence the vapour release, adding their own specific odour to the target scent, so we need to be careful with our choices.

If we place a narcotic, such as cocaine, on a surface, that surface will smell like cocaine, even when the powder is removed. Odour carriers and ODDs will prevent such contamination taking place.

Traditionally, professional detection dog handlers in police and military were convinced that, when training a target odour, they had to work with the pure basic ingredients of explosives narcotics – the real stuff. However, we have now learnt the dogs need to be trained also on a mixture of odours. This is because, in the real world, the target odour is rarely presented in pure form. In most cases, it will be found in a mixture with other odours. Training detection dogs on mixtures helps them become adept at recognising target odour among other odours, which makes them more effective in real-life operations. Its is also more cost effective to train scent detection dogs on mixtures, as the cost of pure odour samples is higher than mixtures.

There is also on-going debate surrounding the use of real odours, pseudo odours and other options, such as a tennis ball or a piece of kong when a dog first starts training. To clarify:

- Real odour refers to using the real, original material as the target odour. So, for example, if we are training a dog to detect drugs, we would use real narcotics, such as cocaine.
- Pseudo odour makes use of a substance/scent, which replicates, as closely as possible, the real material. By adding different ingredients to the original substance, or by using different dilutions it will mimic the target odour. Pseudo odours are used in scenarios where trainers cannot use the real material because it is unavailable or too costly, or because it is illegal, or too dangerous. Also known as parent or simulation odours, they are almost the same as the real material – but not 100 per cent the same.

This begs the question, what is 100 per cent real target odour? Every manufacture will give their own special 'signature' to the 100 per cent target material that is produced. Cocaine is cocaine, and we have found that dogs will use the methylbenzoate chemical (the basic component in cocaine) to identify, find and alert on it. But if we test cocaine from 10 different legal and illegal producers, we will find 10 different signatures. This also applies to explosives. The TNT explosive material made in USA will have a different 'signature' than the TNT produced in Europe.

So whatever target odour is used to imprint and train a dog, we must ensure that training is generalised so he is able to detect multiple different samples of the same target odour. Depending on the goal in training, this may involve getting samples from across the globe. If this is not possible due to legal or shipping restrictions, absorbent materials, such as GetXent tubes, Whatman paper, a piece of cotton or Calico fabric can be used, or whatever material is available and affordable.

There are also options with regard to extracting the scent from cocaine, or other odours; it can be diluted or trapped, allowing us to work with very small concentrations. It is important to think, plan and do before training gets underway, especially when we need to choose a 'training' target scent. Remember, the key to success is generalisation.

The big question is how will these different methods, used in basic training, affect dogs as they progress to operational training and later to real life scenarios where 'finds' need to be produced as evidence in court. As yet, there is no definitive answer and research is on-going. For example, we don't know if dilution produces better results than the original odours and, if this is the case, what would be the best dilution factor to use to train dogs for real-world operations? Better insights into this could also help us to develop a safe system for training with dangerous substances, such as the powerful anaesthetic fentanyl or the highly explosive TATP, and many others dangerous chemicals.

SUMMING UP

When it comes to animal training, I believe we should follow the science and listen to what the scientists are telling us. But we also need to trust our own instincts. Working with animals requires self-belief, emotional intelligence, and empathy. We need to be curious, keep an open mind and look beyond the limits of our own truths.

I have had the immense good fortune to learn from many different people, and from many different animals. Albert Einstein once said: *"If you can't explain it simply, you don't understand it well enough"* – and those words have always rung true for me. Look for people who can explain simply. They understand what really matters and can help us grow both as animal trainers, and as people, so that we can, in turn, offer support to others.

In writing this book, my aim has been to present challenges and to provide motivation so that guardians, handlers and trainers feel inspired to try new ways of doing things. This is not only to make training more effective and efficient, but also to achieve a greater understanding of how dogs respond, process information and learn new skills. I hope I have succeeded in some part and, added to that, my greatest wish is that we all work together – sharing knowledge and ideas – so that we influence the world of dog training for the benefit of all concerned.